Bodily Boundaries, Sexualised Genders & Medical Discourses

Bodily Boundaries, Sexualised Genders & Medical Discourses

Edited and Introduced by
Marion de Ras and Victoria Grace

The Dunmore Press

First Published in 1997
by
The Dunmore Press Ltd
P.O. Box 5115
Palmerston North
New Zealand

Australian Supplier:
Federation Press
P.O. Box 45
Annandale 2038 NSW
Australia
Ph: (02) 552-2200
Fax: (02) 552-1681

ISBN 0 86469 282 X

Text:	Goudy Old Style 9.5/11.5
Printer:	The Dunmore Printing Company Ltd Palmerston North
Cover art:	Samuel van Hoogstraeten: *The Anaemic Lady*, Rijksmuseum, Amsterdam
Cover design:	Vanessa Halley

Contents

Introduction

'Corpus Mobile': Body, Gender, Sexuality and Medicalisation

Marion de Ras
and Victoria Grace

The body forms one of the most intriguing subjects of research and theory in feminist thinking and critique and in other 'new' academic disciplines such as cultural studies, queer studies, indigenous studies, ethnic studies and post-colonial studies. Feminist books on the body proliferate, not to mention the numerous articles and journals dedicated to the body which have appeared in the last 20 years and the extensive writings of French poststructuralist authors on gender, subjectivity and embodiment.

The new study areas, and feminist studies in particular, have a specific interest in the body. Female bodies, gendered bodies, colonised bodies, coloured bodies, homosexual and lesbian bodies, ethnic bodies, children's bodies and disabled bodies have been researched, defined, spoken about, organised, ritualised, regulated, disciplined, enculturated, objectified and thus 'constructed' as 'other' in the long-term discourses and praxis of natural and medical sciences and of Western thinking and philosophy in general.

This interest has, over the past 20 years, developed into a multifarious 'body' of topics and an equally diverse 'body' of theoretical perspectives, epistemological standpoints and methodologies. Radical feminism, with its urgent concern to challenge and oppose the gendered body politics of the patriarchal society, made a beginning focusing foremost on the objectification, thus on the exploitation and oppression, of the female body, in particular in relation to sexuality and reproduction. Feminist critical theory, combining anthropological, sociological and historical perspectives, analysed the complex processes of categorisation and construction, emphasising the eighteenth and nineteenth century Western scientific constructions of binary oppositions such as man/woman, male/female, masculinity/femininity, nature/culture, sex/gender, body/soul, body/mind, and pointed to the inherent contradictions of 'man is culture' and 'woman is nature'. The view that 'any linkup between woman and nature is a construct of culture rather than a fact of nature' has guided generations of scholars into researching 'gender' as a social construction. Postmodern psychoanalytic feminist theory perceived the female body and gender within the construction of the (never complete) Subject and its

Desire, whilst poststructuralism and postmodern theory in general declared the death of The Body as an ontological category, essence or material, focusing rather on that which is, according to these scholars, very much 'alive': the multi-faceted forms, images, styles and performances of the body, gender and sexuality, its constructed boundaries and borders and the processes of passage, crossing and transformation. According to Judith Butler, gender, for example, is not an essential quality or category, nor a 'material' reality, but a performative action which needs to repeat itself over and over again in order to be perceived, understood and conceived as material.

Over the last 20 years, diversities of theoretical perspectives, epistemological positions and methodologies, have resulted in sometimes insurmountable antagonisms within feminist theory and practice, leading to the creation of new dichotomies such as essentialism versus construction, socialisation theory versus postmodernism, materiality versus representation. Despite these antagonisms, however, there is a substantial stream within feminist academe that takes as its starting point the statement that we cannot speak of the body, gender and sexuality as fixed entities, nor can we speak of a distinction between the body as natural and gender as cultural, or a distinction between the body as the site of biology and gender as a social construct. The body, gender and sexuality are only meaningful within language and are thus influenced by history, culture, the social and political, religion and philosophy. The scientific perceptions of the body and bodily processes, our perceptions of the body and bodily processes, our perceptions of our own body are influenced by these grand processes, which are constantly changing and are always culturally regulated. The Body does not exist outside language, outside culture, outside history, nor does our perception of the 'materiality' of the body.

Michel Foucault's historical analysis has proven to be of paramount importance for feminist thinking and research focusing on the medicalisation of the female body. Many of the authors in this book refer to his theory. His analysis of Scientia Sexualis as a Western process, as a process of a 'will to knowledge' and a knowledge/power/sexuality-nexus, indeed has led generations of scholars into the research of the discourse of sexuality, the regulation and disciplining of bodies and the process of medicalisation. His description of the four grand strategies of the regulation of sexuality, body and gender: the hysterisation of the female body, the pedagogisation of the child's sexuality, the socialisation of procreative behaviour and the psychiatrisation of perverse pleasure, has influenced feminist scholars to perceive the body both as a medium of culture and a locus of control and led them to research the constructions of femininity as texts and practices of regulation, normalisation and discipline. Within this framework the historical constructions of body, gender and sexuality are understood as political, and authors influenced by Foucault, such as Susan Bordo analysing the phenomenon of female disorders, will state accordingly: whether we look at hysteria, agoraphobia, or anorexia, we find the body of the sufferer deeply inscribed with an ideological construction of femininity.

This book is a result of a conference on Body, Gender, Sexuality and Medicalisation, held in May 1995 at the University of Waikato. The authors of the chapters in this book focus on a variety of feminist concerns, endeavouring to understand the body as socially, culturally and historically constructed. The majority of the contributions are, in various ways, concerned with a domain of enquiry where medical discourses and practices intersect with experiential discourses, socio-political analyses and cultural representations.

If we are to contemplate the body as socially constructed, then the focus of attention will turn to interpretation of the discourses that contribute to this construction. We will want to ask – what social meanings inscribe the body? What representations or signifying systems articulate bodily experience? What strategies of power are involved regarding gender and sexuality? How might we 'read' the body as a social and cultural text? How can we understand the social practices that situate and circulate these representations? And simultaneously, what do we make of our own actual bodily experiences? What constructions, what 'architecture' of body, gender and sexuality emerge and how are they to be related to long-term processes, especially regarding the history of medicine and the process of medicalisation.

In critiquing the empiricist assumptions of objectivity in the way (bio)medical science constructs (female) bodies and to focus on these constructs as representations, we recognise the epistemological problems which have recently surfaced among feminist scholars regarding the notions of materiality of the body and cultural constructions or representation of the body. Urgent questions have been asked such as: if bodies are always representations, always cultural constructions rather than natural/biological facts, how then do we approach the matter of materiality and how do we approach the matter of this new dichotomy, materiality versus representation, which we seem to have constructed ourselves?

However, by adopting an assumption that knowledges are inevitably discursive, in other words we do not have direct access to 'reality' but rather our representations of reality are unavoidably socially constructed (which means that reality is socially constructed), this does not mean that we assume that reality is reduced to discourse. Taking a critical epistemological standpoint means we problematise the notion of neutral observers uncovering universal laws of nature through scientific method. It is important to stress that this critical stance does not negate the value of, for example, biomedical research. Our focus is directed at the re-reading of scientific texts and practices as discursive and as culturally, and historically, situated. It is this critical re-reading that has attracted us to the chapters selected for this book. The contributions that follow, work, in the main, by creating dialogues across discursive fields. In doing so they throw light on the complexity of the social processes at play in constructing knowledge.

Medical discourses and practices are thus understood here as productive of bodily representations in a way that is not positioned against body constructs of the 'natural' and 'normal'. Rather both of these discursive fields are interdependent and mutually problematic in their assumptions. Medicalisation, therefore, can be understood as the historically-specific social process whereby medical discourses play an important role in constructing culturally powerful significations of bodies and creating the institutional structures for medical regulation and intervention.

Bodily Metaphors, the Politics of Bodily Representation and 'Reading' the Body as 'Text'

Throughout the history of the construction of femininity, the female body has, in Western philosophy and scientific discourse, served as a signifier of a variety of social categories: nature, health, regularity, strength, stability, nationalism, land, the rural, the machine, the industrial model. At the same time the female body in Western and non-Western societies, through its bodily emissions of menstruation, sexual fluids and lactation, also has been represented to incorporate the signs of bodily, gender and sexual pollution. Whilst the female body is put on display to represent nature, beauty, unity, health, moral virtue and regularity, signalling the 'perfect' body, it is also is perceived as the imperfect, the 'uncivilised' body: inhibiting as well as creating disease, weakness, uncontrollability, irregularity and abnormality; the unhygienic body.

In contrast the male body in Western and non-Western societies has been, metaphorically speaking, perceived as the norm, as the category which needs no questioning and no research into its 'maleness'. The male body is perceived and thus constructed as the One, as 'whole', as 'closed'. The female body is, in its essence of femaleness, perceived as the Other, as 'open', as 'leaky', as possessing too many orifices. Because of this ambiguity, which is ultimately the result of perceiving the female body from the assumption of the nature/culture dichotomy, the obstinate female body with its incomprehensible gender and sexuality, therefore has been regarded as an entity or category which needs controlling, purifying and disciplining. The deployment of metaphors of female bodies and of female bodily processes then functions to construct difference and to allay the danger of blurring and crossing of geographical, sexual, bodily and gender boundaries. For the blurring of boundaries questions the fixity of anything that presents itself as One, ultimately invoking change and thus influencing the processes and strategies of power.

The authors in the first part of this book, *Bodily Metaphors,* are specifically concerned with political, cultural, geographical and/or scientific-medical metaphors, with metaphoric representations of the body and with the ways the body and bodily processes are viewed, understood, focused on, defined and 'read'.

Emily Martin's chapter *The New Culture of Health: Gender and the Immune System in America* concerns the revealing and opening up of so-called 'dead' metaphors in scientific, especially medical, language in order for it to become more receptive to its own interrogation. Those readers acquainted with Martin's work will know that her former project *The Woman in the Body* was concerned with the analysis of medical and scientific metaphoric representations of female reproductive processes such as menstruation, birth and menopause and how these became understood as and compared to images of a functioning or malfunctioning industrial model. In this article her focus is on the recent powerful imagery of the 'healthy' body as, influenced by AIDS, it has become related to the modern image of a strong immune system. Martin analyses the wider social and gender implications and connections of such 'bodily' requirements: productive, flexible, with responsive immune systems. The 'flexible body' signals specific positive gender images as well as new images of the modern worker: fast, mobile, healthy, flexible, in a modern globalised industry. Critically examining this new imagery, Martin questions this 'alluring model' for its risks.

Pamela Wood's project *Body Boundaries: Using the Metaphor of Body in Understanding Public Health Practice in Nineteenth Century Dunedin*, focuses on nineteenth century Dunedin, the debates and organisations and regulations of hygiene, analysing these in terms of the metaphor of the city-body. She uses the metaphor of the 'body' and in particular 'body boundaries' for exploring public health problems and responses in nineteenth century Dunedin. The emerging settlement is regarded as a 'city-body', thrown out of equilibrium by the sudden and dramatic rise in population with the Otago goldrush. Sanitarian concerns focused on topographical boundaries such as swamps and the marshy margins of the town and on the dirt, mud, rotting waste and excrement produced by the rapidly growing society. The city-body's main problem was one of containment, represented especially by its overflowing cesspools, stinking open drains and stagnant bogs, making it a leaky, swampy body, seeping from every orifice.

Judith Macdonald's *The Body of the Land – The Bodies of the People: Gender in Tikopia*, concerns the metaphorical connections between gender, body and land: society as a 'body' with boundaries, margins and internal structures. Her chapter describes the ways in which the land is conceptualised as a body and the delineation of internal and external boundaries of that body which are also markers of gender-specific realms. Analysing Tikopia understanding of the physical body and its social articulation in the gendered landscape, Macdonald withdraws from the simple nature/culture framework which posits women's bodies as 'natural' and therefore needing to be brought under control, proposing a framework which allows for a more complex understanding of the metaphorical extension between body and land.

Heather Worth's *AIDS, Race and Gender* focuses, again in a metaphorical sense, on 'jungle fever'. The imagery of the black African body infecting New Zealand ('pure') bodies. AIDS has changed the way we think about the human body and its capacity for sexual pleasure. It has become the site of competing struggles around sexuality, gender and race and has engendered Western-centric debates around immigration and has intensified irrational fears about links between outsiders and HIV infection. Peter Mwai, a Kenyan musician, was convicted in Auckland, New Zealand of a number of offences relating to infection and possible infection with the HIV virus. Worth's chapter argues that the xenophobic response to Mwai, as African, obscured the real issues around women, safe sex and HIV.

Where the chapters grouped together in the first part of the book explore and elaborate the interconnections between meanings of the body and meanings of the wider social, political, geographic, judicial and racial landscape through metaphorical readings, the chapters in the second section, *the Politics of Bodily Representation*, examine the political investments contributing to particular forms of bodily representation. The focus is specifically on *gendered* political investments and bodily representations that affect girls and women. Authors consider questions arising from the silencing of some discourses and the privileging of others, in relation to menstruation and pelvic pain. What gets said and why, how bodies are represented, is shown, in both of these examples, to be intensely political. Similarly, the medical, dieting and fashion industries' investments in representations of bodily size is examined and how women who don't fit the standard bodily dimensions are implicated and located within this powerful social ideology. The focus on the historical and contemporary explanations for, and constructions of, 'maiden diseases', recently anorexia nervosa and bulimia nervosa, argues for the need to question the epidemic of discourse on the representation of the bodies of girls and young women as inherently 'vulnerable'.

Kathryn Lovering, in her chapter *Listening to Girls' 'Voice' and Silence: The Problematics of the Menarcheal Body*, situates her research, on the inscription of adolescents into discourses of menstruation, in an historical context of discourses constituting female puberty. She is concerned to analyse the process of production of 'shame and bleeding that is beyond common language' and to understand the production of this shame and embarrassment in terms of the significance of menstruation as a marker, or *the* marker, of sexual difference. Lovering's analysis is based on research recently completed in Britain exploring girls' talk about menstruation. She details the way girls locate themselves within discourses of menstruation, the subject positions they take up and those precisely absent through the politics constructing representations of the menarcheal body. Her research shows the intransigence of dominant discourses of the female body and their resistance to any fundamental change. Lovering does, however, use historical evidence of non-pathological constructions of menstruation as well as material from her research transcripts, to suggest the importance of different kinds of talk and practices surrounding menstruation to subvert those discourses positioning the female body as 'other'.

Victoria Grace in *Reading the Silent Body: Women, Doctors and Pelvic Pain*, explores the medical consultation for chronic pelvic pain. She examines the implications of the medical reduction of pain to the physiological and the specular which leads too frequently to the diagnostic assertion that 'nothing was found'. This assertion is typically accompanied by an unspoken spectre of 'psychological problems' which neither the gynaecologist nor the psychiatrist seem to know anything much about. Grace develops a theoretical analysis of the implications of this ontological dominance of the specular within the mind/body dualism of medicine and its implicit gendering. She is concerned to develop an understanding of the doctor's exercise of power in the consultation when confronted with that which is 'unrepresentable' in (bio)medical terms and develops this analysis through exploring connections between the imposition of bodily pain and the destruction of language and 'self'. She argues that we need a new approach to *practising* an understanding of chronic pelvic pain.

Jenny Carryer's discussion of the culturally-dominant, medicalised construct of 'obesity' suggests that, in the experience of 'embodied largeness', the politics of the discursive representation of the female body excludes, in almost absolute terms, the possibility of a competing discourse, a resistant subjectivity that accepts and enjoys one's own female body as 'large'. Carryer introduces her topic, *The Embodied Experience of Largeness: A Feminist Exploration*, by situating discourses on body size within the dominant medical frame of health and illness, normality and pathology. Disputing conventional medical wisdom concerning the need to ensure one's body fits within a narrow range of acceptable sizes to comply with health edicts, Carryer highlights the influence of medical discourses in creating and sustaining the association of bodily largeness with profound (Western) culturally-inscribed, negative moral sanction. She then moves to a consideration of the implications of cultural imperatives regarding body size, gender and femininity and prefaces a discussion of her research participants' narratives with a feminist critique of discourses on women's bodily size. The narratives of her research participants are unambivalent in their marking and re-marking a position as women subjected to an insistence that they battle their 'unruly bodies'. Feelings of shame are described by Lovering as associated, albeit ambivalently, with the menarcheal body. Shame appears powerfully in the experience of the women with large bodies interviewed by Carryer.

Marion de Ras' project *'Maiden diseases' and the Pathologising of Female Puberty and Adolescence* concerns eating disorders such as anorexia nervosa and bulimia nervosa and girls' illnesses in general. Her interest lies not with exploring the etiology of girls' illnesses. Rather, from a socio-historical and cultural framework, she is interested in the process through which certain constitutions of girls and young women, in the course of Western history, became perceived as 'female' and belonging to a specific life phase, namely puberty and adolescence and diagnosed as illness. According to de Ras the assumed incidence of anorexia nervosa and recently of bulimia nervosa reflects more than just the history of an illness. It reflects a history of focusing on and problematising sexual, bodily, social and moral phenomena of girlhood. Her concern is the process of 'focusing on', or 'gazing at', together with what is, in her view, the subsequent product of such focusing on: the construction of girls' diseases.

The Western scientific process of rationalisation, of 'wanting to know' the body and in particular the female body, gender and sexuality, went hand in hand with the praxis of visualisation, of the 'scopic', of looking at and opening up the dead and the living body, perceiving its outer and its inner, its surface and depth, searching for and understanding its 'truth' so as to ultimately define, categorise, catalogue and systematise this knowledge. This process we call, following Foucault, the 'writing' of the 'truth' of gender and sexuality on the human body, of femaleness on the female body. Authors in the last section, *Reading the Body as Text*, are concerned with unravelling such 'knowledge', such processes of 'inscription', by way of an analytic process of deconstructing the 'gaze' and by way of art historical and iconographical perspectives considering bodies put on 'display', in a variety of (combinations of) settings: the autopsy table, a painting, the critical care unit.

Susan Duncan's *Autopsy as Gaze: The Construction of the Hermaphroditic Corpse as a Text of Sexual Difference* analyses the medical gaze as both visual instrument ('reading') with a view to 'know' about the object – the 'true' sexuality and gender of hermaphrodites through dissection – and instrument in ('writing') creating such knowledge – the ('textual') construction of sex and gender in the dissecting process, through the process of classification and definition. This process, according to Duncan, enacts the creation of difference. She argues that there is always a gaze behind anatomical inscription and description. In this respect, the male medical gaze is one that seeks to define the heterosexualised duality of male/female as subject in relation to that which is different and separable as an object; the hermaphroditic body.

Heleen Buijs' *The Imperfect Woman: The Female Figure in Dürer's Melencolia I* analyses the depiction of melancholy as a woman. She questions the phenomenon that, although melancholy is depicted by a female figure, this has never been a serious object of research in the vast literary production of art historical interpretations of one of Dürer's masterpieces: *Melencolia I*. Art historians who have incorporated the femaleness of the allegory have done so from an ahistorical and traditional view of women in general. Buijs introduces a perspective which centralises the female figure, interpreting her as allegory as well as woman, questioning why melancholy is represented through a female figure, placing it against the historic and philosophic background and in relation to Dürer's late medieval and early Renaissance understanding of, and ambivalence towards, the 'new woman'.

As Mary Morrison states in her introduction to *Body Guarded: The Social Aesthetics of Critical Care,* the genesis of her project is founded in negotiating a terrain that incorporates the two disciplines in which she is trained: fine arts and critical care nursing. Questioning the disparity of these worlds – both involve the human condition and require observation and visualisation skills – Morrison evaluates critical care artistically, using a theoretical as well as a visual approach. Body-guarded refers to the protective needs of incapacitated patients, and social aesthetics to the use of art to examine social aspects of critical illness. Medicine uses the justification of medical science to conduct research into the pathology of the critically ill, but Morrison believes that it is also valid to evaluate that human experience using fine art. This project, then, allows from a fine arts perspective a critical visualisation and creative challenge of medical treatments of the critically ill, medical models and medical metaphors such as the body as machine.

To conclude, we would like, in a metaphorical and epistemological sense, to call the project of this book 'corpus mobile'. Corpus mobile, refers to the variety and diversity of 'body' themes in this book, to the theoretical perspective that bodies are both object and subject in their 'being', expressions and representations, thus transgressive and boundary crossing, thus 'mobile' and it refers to the theoretical perspectives themselves which, if we perceive these as 'bodies of knowledge and critique' are (or at least we hope they are) mobile in the sense of transgressing disciplines, analysis, standpoints and perspectives.

Part 1

Bodily Metaphors

The New Culture of Health

Gender and the Immune System in America

Emily Martin

The science of reproductive biology includes in its inner core, its very language and concepts, deeply cultural assumptions about males and females. In this chapter I will analyse some words and images from medical and popular representations of women's bodies, with an eye to their imagery and an ear to their poetry. Key to this task will be 'waking up' or bringing to life some of the overly familiar and therefore unseen metaphors ('dead' metaphors as students are taught to call them in high school English classes) in scientific language. As long as they are 'dead', Toni Morrison's words describe them:

> A dead language is unyielding language content to admire its own paralysis. ... Unreceptive to its own interrogation, it cannot form or tolerate new ideas, shape other thoughts, tell another story, fill baffling silences.
>
> (1994: 13–14)

If, in a productive synergy between biomedicine and the anthropology of science, dead scientific language could be enlivened, perhaps the result would be a scientific language of the body that was receptive to its own interrogation, tolerant of new ideas and other thoughts, open to other stories and able to acknowledge the existence of (if not fill) 'baffling silences'.

The Body as Machine

In earlier work I have argued that the standard medical accounts of a woman going through menstruation, birth and menopause depict her body as metaphorically engaged in various forms of production on an industrial model. When she menstruates instead of getting pregnant it is interpreted as a result of failed production. (Menstrual fluids, which one author of a standard text used in medical schools described as 'the uterus crying for want of a baby', are seen negatively, as the result of breakdown, decay, necrosis or death of tissue.) When she gives birth it is regarded as successful production, but this production is often held to a rather strict timetable, reminiscent of assembly-line production. When a woman reaches menopause, the main headquarters governing her body's production system is thought to undergo a devastating breakdown leading to loss of the centralised control necessary to keep order (Martin, 1987).

I have also argued that the standard accounts of fertilisation (in medicine as well as popular science) see the egg metaphorically as a damsel in waiting, or a damsel in need of rescue and the sperm as her seducer or rescuer (Martin, 1990). In a drawing on the cover of a technical biology journal, the egg is depicted as a kind of sleeping beauty, startled awake by the sperm's kiss. A popular cartoon shows the egg as a housewife, besieged by sperm who have devised clever disguises to trick her into opening the door for them.

All this imagery depicts a body that belongs to the late industrial machine age, especially the post-World War II era of mass production, dominated by orderly assembly-line production on a rigid time schedule, run by machinery divided into parts, each with a separate function. Even the courtship drama of the egg and the sperm is rigidly patterned on a post-World War II model, machine-like in its inexorable conclusion in heterosexual union and fecundity.

In its machine-like solidarity and concreteness, this body has definite edges. Where the body ends and the outside world begins is marked with precision and is emphasised because this border is where the battle for health goes on. The 'Castle of Health' illustrated home health books in the 1920s and 1930s and shows how important the outer walls of the body's defence were. In this drawing, although the body is shown as a castle defended by no less than two walls and moats, we cannot see the interior of the body-castle at all (Ritchie, 1918: 16). The 'Lilliputian Hoards' from 1955 shows the preoccupation at the time with body surfaces: all the action is outside the body on its surface, the skin, where demonic little germs are trying to puncture the barrier of the skin with sharp objects like drill bits (Coughlin, 1955: 122). And of course this concern with the integrity of surfaces went along with the preoccupation of the times with hygiene, especially cleaning and protecting the outer surfaces of the body, as well as almost obsessive washing, dusting, airing and disinfecting the surfaces of the home or the community.

I have spoken of this body of clear boundaries and machine-tooled parts as if it only lived in the 1950s. In fact, it is very much with us still. To mention just one domain, contemporary accounts of the immune system in both research and popular contexts see it as maintaining the health of the body through continuous warfare against the foreign enemy. This scenario presupposes an absolute difference and a clear way of defining what is friendly 'self' and what is unfriendly 'non-self'. Imagery of aggressive immuno-warfare against the foreign, focuses on a body that is all of one kind, all purely self. This pure body is the 'normal' body, the desirable body to have. Once again, it is as if the body were a castle and its ramparts held stalwartly against anything foreign ever entering.

As Donna Haraway puts it, 'The perfection of the fully defended, "victorious" self is a chilling fantasy' (Haraway, 1992: 320). She asks, 'when is a self enough of a self that its boundaries become central to institutionalised discourses in biomedicine, war and business?' (p. 320). Looking at the cover of a popular book about the immune system *The Body Victorious* by Lennart Nilsson, gives us very specific clues about the identity of this self, in terms of both race and gender (Nilsson, 1985). The cover shows a nude muscular male figure, back-lit so that the edges of his body are sharply outlined. Enough stray light escapes to identify him by his facial features and hair type as white. This is part of the cultural construction of whiteness, the unmarked category. Superimposed across his abdomen is a greatly enlarged colour photograph of an immune system cell, a macrophage, in the act of eating up bacteria. This is a body whose boundaries are defined extremely clearly. Inside is only self; outside is only non-self. Should any foreign matter enter, it will be swiftly dispatched by the roving armies of the man's immune system.

Compared to the internal purity of this body and self, women's bodies and selves fall far short. When they are pregnant, they are truly hybrid, uneasily 'tolerating' the foreign fetus deep inside. Technical immunological articles wonder 'Why does the mother not mount an attack against the fetus' every cell of which is marked immunologically as non-self. An article in *The Economist* wonders 'Why Does the Body Allow Fetuses to Live?' (1985).

In addition to the 'mixing' of self and other in pregnancy, women are statistically more prone to auto-immune diseases. These diseases are conceptualised as caused by the immune system mistakenly attacking self. In one illustration of a woman with Lupus, an auto-immune disease, she is shown lying inside the ramparts of a castle (*The Sun*, Baltimore, MD, 3 Oct., 1989). But instead of protecting her against threats to her health, the sharp spikes on top of the castle walls are turned inward: the castle of her body has literally turned against her. The depiction of a woman in this illustration is no accident: the American Auto-immune Related Disease Association estimates there are 50 million Americans affected by 80 known auto-immune diseases and most of them are women (Brody, 1994). On every count, then, women lose when measured by this sharply bounded ideal body: they mix self and non-self inside and they frequently mistake self for non-self.

The Body as Complex System

A set of metaphors with implications quite different from those that construe the body as a sharp-edged machine engaged in centralised production is currently exercising a significant amount of influence in some medical specialties. Scientists are deploying metaphors derived from chaos theory, also known as non-linear dynamics or complexity theory. Cardiologists, for example, are coming to see the heart, not as the quintessential mechanical body part, the pump, but as a self-organising system that only beats with a mechanical regularity when the body is near death.

Until recently, it was widely held that sudden cardiac death represented an abrupt change from the apparently periodic state of the normal heartbeat to one in which chaotic arrhythmias occur. Work from several sources has suggested that under *normal* conditions the heart has chaotic dynamics and that near fatal disturbances of the cardiac rhythm are often preceded by a *decrease* in the degree of physiological chaos. This represents a reversal in the conventional usage of the term 'chaos' when applied to the injured heart (Skinner, 1990: 1019). (See also Denton *et al.*, 1990.)

What would happen if, as a thought experiment, we described some of women's reproductive functions, such as menstruation and menopause, in terms of a chaos model, leaving aside the body as mechanical factory or centralised production system?

Our thought experiment might pose the question: have the periodic *regularities* of the female hormonal and bleeding cycles between puberty and menopause been overemphasised just as the regularities of the heart beat have been? In the current model, regular periodicity between well-defined limits is considered normal-estrogen, progesterone and other hormones are produced (if all is normal) with machine-like regularity; menstruation occurs (if all is normal) with the periodicity of a metronome. Disease produces irregularity and shifts between stages of maturation (puberty and early menopause) produce irregularity. Regularity is normal, good and valued, irregularity is abnormal and negatively valued.

Menstrual irregularity is often regarded medically as a pathology related to some organic dysfunction. The dysfunction is variously attributed to a 'presumed' malfunction of the ovaries, or such problems as hyperandrogenism (Arai, 1994), diabetes (Adcock *et al.*, 1994), PMS (Khella, 1992), or anorexia nervosa (Whitaker, 1992). I do not mean to suggest that these correlations are spurious. Rather I want to call attention to the unexamined assumptions that normal equals periodically regular. There is a sharp contrast set up in the typology of the normal and regular vs. the abnormal and irregular. To begin to move toward a different view, one might ask, *how* regular are most women?

I know of no ethnographic study of the topic, but it seems very likely that women experiencing irregularity in their periods in puberty or during early menopause will be disturbed and anxious about the irregularity itself. If the body is a machine, then a faltering or erratic pattern of behaviour seems to portend its imminent failure. Women who are deemed irregular may be given medication to produce regular periods (epidemiologically, 'correction' of irregularity is deemed a short-term benefit of oral contraception [Runnebaum, 1992]). Women who are uncomfortable with their perceived irregularity may demand relief.

In the machine model, regularity is a sign of health, irregularity a sign of disease or impending death. In the chaos model, it is just the opposite.

The conventional wisdom in medicine holds that disease and ageing arise from stress on an otherwise orderly and machine-like system – that the stress decreases order by provoking erratic responses or by upsetting the body's normal periodic rhythms. In the past five years or so we and our colleagues have discovered that the heart and other physiological systems may behave most erratically when they are young and healthy. Counterintuitively, increasingly regular behaviour sometimes accompanies ageing and disease. Irregularity and unpredictability, then, are important features of health (Goldberger *et al.*, 1990: 43–44).

This new picture overturns the earlier notion of the heart as part of a stable homeostatic physiologic system, in which the goal was to reduce variability and 'to maintain a constancy of internal function. According to this theory ... any physiological variable, including heart rate, should return to its "normal" steady state after it has been perturbed. The principle of homeostasis suggests that variations of the heart rate are merely transient responses to a fluctuating environment' (p. 47). In contrast, the new finding suggests that 'the mechanism that controls heart rate may be intrinsically chaotic. In other words, the heart rate may fluctuate considerably even in the absence of fluctuating external stimuli rather than relaxing to a homeostatic, steady state' (p. 47).

The reason chaotic organisation might be an advantage to the heart is this: chaotic systems operate under a wide range of conditions and are therefore adaptable and flexible. This plasticity allows systems to cope with the exigencies of an unpredictable and changing environment (p. 49).

If we were to rethink menstruation and menopause in the logic of chaos, here are some possible shifts that could occur. We could describe 'irregularity' as an adaptive response to a changing internal and external environment. The young woman whose menstrual cycle is affected by exercise, by stress or by puberty could then think what a good job her endocrine system is doing flexibly adjusting to her life, rather than worry unduly about a pathological 'irregularity'. Epidemiological studies that have shown greater menstrual irregularity in women who work at night (Miyauchi *et al.*, 1992) and women who are vegetarians (Lloyd, 1991) could be taken to reveal the responsiveness of these women's physiological systems to their particular environment, rather than a pathological deviation from a putative norm of machine-like periodicity. The older woman whose menstrual cycle is affected by approaching menopause or other aspects of her life could do the same. Here it would be particularly gratifying to see irregularity as a sign of vigour and health instead of impending disease and death.

The change women undergo during menopause itself could be described as a phase change, of the sort complex systems often undergo. 'Complex systems sometimes behave discontinuously. Systems simply change states. In models of complex systems, a controlling loop may reach a threshold state and transfer control to another loop altogether. The system appears to have experienced a discontinuous jump from one set of apparent relationships dominating the action to another' (Arney, 1991: 51). For a woman undergoing menopause, thinking of it as a state change from reproductivity to maintenance of non-reproductivity would constitute a far more positive view of her body than thinking of it as breakdown of centralised control.

These suggestions are actually grounded in far more than the academic speciality of cardiology. They are grounded in ethnographic fieldwork I have done in a wide variety of settings – an immunology lab, clinics, a hospice for AIDS patients, support groups, urban neighbourhoods and workplaces – which has convinced me that we are in the midst of a profound sea-change in how the body is conceptualised.[1] Ordinary people in all walks of life, my research has shown, quite commonly see their bodies as complex non-linear systems. This has happened in part by way of an enormous cultural emphasis on the immune system, which has moved to the very centre of the way ordinary people now think of health.

Let me illustrate with an example of how people talk about the immune system:

> [Two men talking, Bill Walters and Peter Herman[2]] *Bill*: I don't even think about the heart anymore, I think about the immune system as being the major thing that's keeping the heart going in the first place and now that I think about it I would have to say, yeah the immune system is really ... important ... and the immune system isn't even a vital organ, it's just an act, you know? *Peter*: It's like a complete network ... if one thing fails, I mean – if ... *Bill*: ... if something goes wrong, the immune system fixes it, it's like a back-up system. It's a perfect balance. The immune system

> is the whole body, it's not just the lungs or the abdomen, it's, I mean – if I cut myself, doesn't my immune system start to work right away to prevent infection? So it's in your finger, I mean – it's everywhere.
>
> (Steven Baker)

From a mechanical body made up of simple components with different functions, these people have moved to a body based on a dispersed, fluid system. Like a fractal the protective ordering functions of the immune system are present in every part of the body no matter how different they appear. One woman we interviewed explained her conception of the immune system this way:

> My visualisation would be almost tides or something ... the forces, you know, the ebbs and flows.
> [Could you draw anything like that?]
> I could.
> [What is it that ebbs and flows?]
> The two forces, I mean, the forces ... imbalance and balance.
>
> (Vera Michaels)

As she spoke, she drew the accompanying illustrations, labelling it 'the waves'. She was reaching for a way of imaging a fluid, ever-changing body; a body containing turbulence and instability, in constant motion; a body that is the antithesis of a rigid, mechanical set of parts. This new body is also in delicate relationship to its environment, a complex system nested in an infinite series of other complex systems.

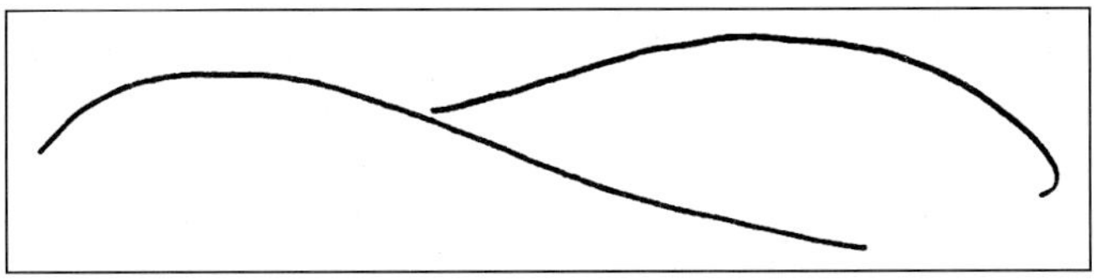

Vera Michael's drawing of the immune system.

It is not only men and women who are not scientists who are working with a conception of a fluid, flexible, ever-changing body. Although they are not yet in the mainstream, a subset of immunologists favour what they call a network theory of the immune system. Imagery of the dance replaces traditional imagery of the immune system in battle against external foes. The body positively reaches out into the world and takes it in:

> The dance of the immune system and the body is the key to the alternative view proposed here, since it is this dance that allows the body to have a changing and plastic identity throughout its life and its multiple encounters. Now the establishment of the system's identity is a positive task and not a reaction against antigens (foreign substances).
>
> (Varela and Coutinho, 1991: 251)

In this new kind of body, attuned to its environment, what counts for maintaining health is what goes on inside. Thus, in striking contrast to the sharp-edged, closed images of the body in the 1950s and earlier, contemporary images more and more show us a body in motion, not in repose and a body with no skin at all, exposing the inner workings of our protective system. A recent cover of *Science* magazine depicts a body whose protective skin has been stripped away to reveal gleaming white lymph nodes under the arms and in the groin, key places where immune system cells are trained and mobilised.

Apart from images of the body produced by immunologists and by non-scientists alike, my research has also shown that flexibility, adaptability and the ability to rapidly change in response to an ever-changing environment with agility and grace are becoming ideal characteristics of work organisations, the government and educational institutions, among others. 'Flexibility' is enjoined of workers, managers, teachers and recent PhDs: all are frequently exhorted to learn how to agilely shift from job to job or career to career as demanded by the constantly changing market. Flexibility is used to characterise the most desirable personality, the highest form of intelligence and the species most likely to survive. It labels countless products and concepts from Nordic Flex Gold, an exercise machine, to flex space, an architectural design.

Returning to the thought experiment I undertook, to reimagine the processes of female physiology, we can now see that imagining the female body as a complex system would give women a flexible, adaptable, responsive physical self, one that is well-matched to these new ideal notions of the kind of self – agile, quick, graceful, in constant change, nimbly responsive to alterations in the environment – now thought necessary for survival in the contemporary world.

But renovating female physiology this way does not depend on the musings of an anthropologist. A biologist has recently produced a new interpretation of menstruation that leads in the same direction. In place of the standard medical view of menstruation, that menstruation is the waste-product of a failed conception, composed of debris, dead tissue, scrap, of no use, Berkeley biologist Margie Profet substitutes the argument, which she bases on comparative evolutionary data, that:

> ... menstruation evolved as a mechanism for protecting a female's uterus and fallopian tubes against harmful microbes delivered by incoming sperm ... the uterus is extremely vulnerable to bacteria and viruses that may be hitching a ride on the sperm and menstruation is an aggressive means of preventing infections that could lead to infertility, illness and even death. In menstruation ... the body takes a two-pronged attack against potential interlopers: it sloughs off the outer lining of the uterus, where the pathogens are likely to be lingering and it bathes the area in blood, which carries immune cells to destroy the microbes.
>
> (Angier, 1993)

I have no way of knowing in detail what lies behind the uproar in the scientific community over these claims: letters to the *New York Times* editor assert vehemently that Profet's ideas lack a 'scientific or medical basis' and insist instead that menstruation is likely to promote infections (Grimes *et al.*,

1993). But I would hazard a guess that the uproar has something to do with two powerful reversals Profet's account accomplishes. The first is that sperm, instead of being gallant and virile rescuers of the hapless egg (the 1950s model) are unwitting carriers of harmful pathogens (germs). The second is that women's menstrual flow becomes, instead of useless and disgusting debris, an important part of her flexible and responsive immune system. Menstruation becomes 'not a passive loss of unused uterine lining, but an aggressive way to prevent infection' brought in by sperm (Angier, 1993).

Sperm carry germs; menstrual fluid washes them away. In Profet's reinterpretation, men's substances lose standing at the very same time that women's substances gain. Whether one is concerned about nutrition, exercise, toxins, stress, cancer or AIDS, the health arena is saturated with immune system talk. The immune system had begun to function culturally as the key guarantor of health and the key mark of differential survival for the twenty-first century. No wonder Profet's model of female menstruation as a major site for immune function where men have none, has raised a few hackles!

However, appealing and seductive to our contemporary sensibilities, imagining the body as a flexible, agile, adaptive system is not without its problems. To be flexible is good, but whose good?

Consider the employees of the Fortune 500 corporation where I did fieldwork, who were trained by exercises on the highwire to become flexible, agile and fearless, to 'work without a net' so they could rapidly adjust to continuously changing conditions, especially if (or rather when) they were 'delayered' (fired). We need to realise the cost of being willing to change jobs many times, which may include loss of pensions, health insurance and unemployment insurance. We need to realise in the face of the incitement to be nimble and in constant motion, the common human need for stability, security and stasis.

Consider the ubiquitous ads for everything from communication systems to shoes that enjoin us to become 'Flexible Folk' who will move fluidly through space and time to make ever-changing connections with our environment. Fresh from recent experiences of the body held to rigid postures, strict timetables and limited movements, it is no wonder that moving gracefully as an agile, dancing flexible worker/person/body feels like a liberation even if one is moving across a highwire. But can we simultaneously realise that these new flexible bodies are also highly constrained? They cannot stop moving, they cannot grow stiff and rigid, or they will fall off the 'highwire of life' and die.

Why not just join in the celebration of such an appealing way of being, energised, alert, engaged and active? One reason is because this ideal image is based on a tiny portion of human capacities. It is an ideal some people will be able to achieve better than others. While we are striving for the flexible, lean and agile, let us remember the virtues of the stable, ample and still. A second reason is that even though the new model of the body in constant change and adaptation might allow women's bodies to come closer to the new ideal than men's, I for one am reluctant to embrace uncritically any model that discriminates, even if it would reverse an age-old imbalance. We must resist uncritical adoption of this alluring model, for it risks making us think there is a natural basis why flexible, nimble men and women with responsive immune systems and resilient personalities survive in decent jobs, while others of us, with flawed or rigid immune systems, will slide into poverty and disease.

I began by suggesting that we 'wake up' the dead metaphors in scientific language so it can become more 'receptive to its own interrogation'. The point of waking the language up is not to remove metaphor from scientific descriptions, to sterilise it to abstraction. It is to open our minds to the presence of metaphor, the power of metaphor and to the possibility that scientific language awakened to its cultural powers might become a more 'complex, nuanced midwife' (to paraphrase Toni Morrison) to our contemporary concepts of the self.

Notes

1. The research was funded during 1989–1991 by the Spencer Foundation. Several graduate students collaborated in the research: Bjorn Claeson, Laury Oaks, Wendy Richardson, Karen-Sue Taussig and Monica Schoch-Spana. Some of our findings are described in Martin, 1994.
2. All names of people interviewed are pseudonyms.

References

Adcock, C.J., Perry, L.A., Lindsell, D.R., Taylor, A.M., Jones, J. and Dunger, D.B. (1994), *Diabetic Medicine*, vol. 11, no. 5, pp. 465–470.

Angier, N. (1993), 'Radical New View of Role of Menstruation', *The New York Times*, September, pp. 1, 5.

Arai, K. and Chrousas, G.P. (1994), 'Glucocorticoid Resistance', *Baillieres Clinical Endocrinology and Metabolism*, vol. 8, no. 2, pp. 317–331.

Arney, W.R. (1991), *Experts in the Age of Sytems*, Albuquerque, NM: University of New Mexico Press.

Brody, J.E. (1994), ' "Hair of Dog" Tried as Cure for Autoimmune Disease', *The New York Times*, October, p. C1.

Coughlan, R. (1955), 'Science Moves in on Viruses', *Life*, vol. 38, pp. 122–136.

Denton, T.A., Diamond, G.A., Helfant, R.H., Khan, S. and Karagueuzian, H. (1990), 'Fascinating Rhythm: A Primer on Chaos Theory and its Application to Cardiology', *American Heart Journal*, vol. 120, no. 6, pp. 1419–1440.

Goldberger, A.L., Rigney, D.R. and West, B.J. (1990), 'Chaos and Fractals in Human Physiology', *Scientific American*, vol. 262, no. 2, pp. 42–49.

Grimes, D.A., Padian, N.S. and Washington, A.E. (1993), 'Infection Risk Rises During Menstruation', *The New York Times*, October.

Haraway, D. (1992), 'The Promises of Monsters: A Regenerative Politics for Inappropriate/d Others', in Grossberg, L., Nelson, C. and Treichler, P.A. (eds), *Cultural Studies*, New York: Routledge, pp. 295–337.

Khella, A.K. (1992), 'Epidemiologic Study of Premenstrual Symptoms', *Journal of the Egyptian Public Health Association*, vol. 67, no. 1–2, pp. 109–118.

Lloyd, T., Schaiffer, J.M. and Demers, L.M. (1991), 'Urinary Hormonal Concentrations and Spinal Bone Densities of Premenopausal Vegetarian and Non-vegetarian Women', *American Journal of Clinical Nutrition*, vol. 54, no. 6, pp. 1005–1010.

Martin, E. (1987), *The Woman in the Body: A Cultural Analysis of Reproduction*, Boston, MA: Beacon Press.

Martin, E. (1990), 'The Egg and the Sperm: How Science has Constructed a

Romance Based on Stereotypical Male–Female Roles', *Signs*, vol. 16, no. 3, pp. 485–501.

Martin, E. (1994), *Flexible Bodies: Tracking Immunity in America from the Days of Polio to the Age of AIDS*, Boston, MA: Beacon Press.

Miyauchi, F., Nanjo, K. and Otsuka, K. (1992), 'Effects of Night Shift on Plasma Concentration of Melatonin, LH, FSH and Prolactin and Menstrual Irregularity', *Sangyo Igaku (Japanese Journal of Industrial Health)*, vol. 34, no. 6, pp. 545–550.

Morrison, T. (1994), *The Nobel Lecture in Literature, 1993*, New York: Knopf.

Nilsson, L. (1985), *The Body Victorious*, New York: Delacourt Press.

Ritchie, J.W. (1918), *Primer of Sanitation and Physiology*, Yonkers-on-Hudson, NY: World Book Co.

Runnebaum, B. (1992), 'The Androgenicity of Oral Contraceptives: The Young Patient's Concerns', *International Journal of Fertility*, vol. 37 (Suppl. 4), pp. 211–217.

Skinner, J.E. (1990), 'Neurocardiology: Brain Mechanisms Underlying Fatal Cardial Arrhythmias', *Neurologic Clinics*, vol. 11, no. 2, pp. 325–351.

Varela, F.J. and Coutinho, A. (1991), 'Immunoknowledge: The Immune System as a Learning Process of Somatic Individuation', in Brockman, J. (ed.), *Doing Science: The Reality Club*, New York: Prentice Hall.

Whitaker, A.H. (1992), 'An Epidemiological Study of Anorectic and Bulimic Symptoms in Adolescent Girls: Implications for Pediatricians', *Pediatric Annals*, vol. 21, no. 11, pp. 752–759.

(1985), 'Why Does the Body Allow Foetuses to Live?', *The Economist*, vol. 296, p. 89.

Body Boundaries

Using the Metaphor of Body in Understanding Public Health Practice in Nineteenth Century Dunedin

Pamela Wood

In 1864, 16 years after its establishment as a British colonial settlement in the southern region of New Zealand, Dunedin was still a filthy, muddy city, inadequately drained and without an organised system for sewage removal beyond the services of nightmen. The sudden rise in population in the three years following the 1861 discovery of gold in Otago created appalling sanitary conditions in the town and forced a public inquiry. Citizens complained of the offensive stenches coming from filthy ponds, cesspools and rotting waste – stenches which, according to the prevailing 'miasmatic' theory, were believed to cause disease. Particularly dangerous were the bogs and marshy margins of the town, ambiguous boundaries between water and land, which also produced the miasma threatening public health.

Sanitarian concerns centred on these topographical boundaries and on the problematic animal, vegetable and human waste produced by a rapidly growing society. Although settlers had hoped to impose order on nature, shaping the form of their settlement through planned and steady development, the sudden influx of people to the town threatened its containment and challenged all its boundaries. Tents and shanties spilled down the hillsides, cesspools overflowed and seeped into wells, drains and muddy bogs stank with rotting waste and the very margins of the town merged into the harbour in a swampy mass. Far from being a clearly marked, clean and orderly settlement, the town oozed at its edges and overflowed its channels for water and waste. It was a leaking, swampy body, seeping from every orifice.

The metaphor of 'body' and in particular 'body boundaries' is an appropriate one for understanding public health practice in nineteenth century Dunedin. As Davison (1983) has described, the conception of a city as a social body, an organic or natural system, has been used frequently by social theorists since the late eighteenth century. In this view, the city is akin to a body in that it comprises particular parts, each with its own purpose yet dependent on others for the effective functioning of the whole. As with the body, the city is an organic entity, capable of growth and decay, adaptation and homeostatic adjustment, regulating its functioning to maintain overall social health. When this equilibrium is upset, the city ceases to function smoothly and problems arise.

Like many metaphors, the idea of a city as a body holds within it paradoxical elements as the city is also the body politic, a rational entity, with a distinct form imposed on nature by humans and easily distinguishable from rural environs which were governed far more by nature. This paradoxical element does not diminish or negate the usefulness of the metaphor, it simply brings an intriguing dimension into play.

Miasma, Margins and Dirt in the City-body

For nineteenth century Dunedin, the city-body metaphor offers a useful heuristic device for exploring public health problems and responses, not least because it sits easily with the prevailing assumptions of disease causation and theories of 'medical topography'. From the late eighteenth century, medical men applied to the landscape the same diagnostic approach as they did to the human body. Just as the physician could diagnose disease by casting a discerning eye over the landscape of the human body, identifying the 'humours' of heat, cold, moistness and dryness which were out of balance, so too could they throw their gaze over the natural environment to identify hazards to health. Particular combinations of climate, terrain and vegetation gave rise to sickness. Low-lying, swampy ground generated miasma, noxious vapours which caused disease when inhaled. Absence of any circulation, whether of air or water, was particularly hazardous. In this regard, stagnant swamps enclosed by vegetation, where miasma would lie heavily and accumulate, were especially dangerous. Open land or high ground, where stagnant water was unlikely to lie and where winds could carry away miasma, were far healthier for human habitation. The same notion of miasma and disease alerted people to the dangers of rotting vegetable, animal or human waste, where decomposition created 'putrid emanations' and harmful 'effluvia'.

Foucault (1989) has argued that a significant shift in medical knowledge occurred around the turn of the nineteenth century. Interest in body dissection brought with it not only a new system of classifying and diagnosing diseases, which were now associated with specific lesions and located in particular parts of the body, but also a new mode of treatment and scientific observation sited in the clinic. In a similar way, Davison says, the ideas of medical topography eventually gave way to sanitary science, which used more analytical methods akin to those of the new clinicians and was better able to deal with the new complex problems created by urbanisation (1983). Southwood Smith, Chadwick and others in Britain spearheaded the new sanitary science, characterised by enthusiasm for the effective elimination of waste and the provision of clean water, clean streets and the free circulation of air. Yet in Dunedin, sanitary science was slower to emerge and the old ideas of medical topography held equal sway. After all, both required the effective removal of waste from the city-body, both had an abhorrence of stagnation and foetid decomposition.

In these common concerns lies a consistent notion of dirt, filth and the danger of decay. The idea of 'dirt' is a cultural construct and as such can usefully be explored through the organising frameworks offered by cultural anthropology. Douglas (1975: 50) describes how dirt emerges from a culture's classification system. Cultures create systems of classification to bring order to their world, but the system is always vulnerable at the margins or boundaries between its categories as these are open to challenge. Some things will not fit easily into any category. Others contest the very categories themselves. All of these are dangerous to the system because of their ambiguity. Similarly, things which cross out of their assigned category and move to another also threaten the system at its category margins and are therefore dangerous. Any of these entities 'which blur, smudge, contradict, or otherwise confuse classifications', the culture considers as dirt (1975: 51). In this way, any material, or any person, event or behaviour, which does not easily fit or which challenges the accepted categories of the system is seen as marginal and may be labelled as dirt. Because of its ambiguity, its inability to be placed within the boundaries of a particular category, whatever is marginal is considered not only dirty but physically or socially polluting and dangerous.

Physical dirt may be the result of a breakdown of matter. Because the original matter becomes unrecognisable, as it has lost its initial shape and is now formless, it cannot remain in its originally assigned category. It no longer belongs there. Decomposition moves the substance from one category to another. Mud, for example, is earth shifted from the category of 'solid useful ground' into the category of 'soft, wet, formless dirt'. Rotting vegetables and dead fish, by losing their shape and form, cross out of the category of 'useful food' and become 'unwanted waste' or dirt. In the same way, the concept of dirt as a cultural and symbolic label can be assigned to any person (e.g. leper), occupational group (e.g. prostitutes) or behaviour which contravenes the socially accepted order. Whatever is considered private behaviour, for example, would be inappropriate and considered dirty or dangerous, if demonstrated in a public arena. In all cases, whether a physical entity or a culturally assigned label, dirt threatens order and generates ideas of pollution and social danger which are strongest at the margins of the system. 'Dirt' and 'marginality' are therefore usefully considered together.

Douglas' metaphor of the body incorporates this view of marginality. As with any bounded system, the body is most vulnerable at its internal margins and external openings. Cultures may use the body as a metaphor for their own society in different ways (1973: 16). In the first, the body's head acts as the central control system ensuring effective functioning, with the body seen as a channel for the flow of sustenance and the restoration of strength. The second sees the body as being vulnerable, not from lack of food or rest but from failure to control the quality of what it absorbs through body openings. In this use of the metaphor there is fear of poisoning, a concern for protecting boundaries and an aversion to bodily waste. In a third, rather than having an aversion to waste, a very practical approach is taken to the possible uses of bodily rejects and the recycling of waste.

Each of these is represented in the responses made to public health concerns in Dunedin. The 'rational' view is seen in the policies and regulations made by the local authorities; the 'guardian' view in the concern for clean water, unadulterated food, ventilation, quarantine and the effective removal of waste and sewage; and the 'utilitarian' view in the proposals for the treatment and marketing of excrement. While each has a place in the overall metaphor of city-body, the key metaphoric element is that of body boundaries. It was at the margins and orifices of the city-body that public health danger lurked – the marshy edges of the harbour, the muddy swamps with rotting waste, the open drains and overflowing cesspools – and it was the marginal spaces of the town that were deliberately used for disposal of waste, with the workers themselves considered marginal through their association with dirt. Dirt also allowed internal city boundaries to be crossed, giving public access to the private domestic realm. Whatever was defined as dirt posed a physical or social threat to the settlement and wherever the margins of the settlement were drawn lay the greatest danger of pollution.

Marginal Land

The site of Dunedin was bound by hills, harbour and swamp. In some places the topographical margins were clearly marked by the hills and valleys, but much of the harbour edge was ill-defined, merging into swamp at the northern and southern ends of the town and creating an ambiguous boundary between sea and land. The harbour itself ran north-east for 14 miles, separated from the ocean by a hilly peninsula which rose from the southern marsh. Although the site had natural beauty, it was difficult terrain on which to plan a town as there was little level ground apart from the waterlogged flats. But gradually the settlement took shape, with small rough dwellings put up on the available patches of dry level ground. From time to time congratulatory pieces appeared in the newspapers, optimistic accounts of the progress made in the settlement. Yet other writers complained of the 'stagnant, pestiferous pool, foul and noisome' which the creek in Rattray Street had become and that Dunedin was 'nothing but a swampy, stagnant fen' (*Otago News*, 10 January 1849: 4). By 1855 a Town Board had been appointed to undertake public works, but despite its efforts to form roads and surface them with gravel, the marshes and mud remained consistent features. Residents near the northern swamp were soon used to helping 'Sunday equestrians' who became helplessly bogged, dragging them and their horses out with ropes or pushing planks and beams under them (Fulton, 1922: 132).

Hocken Library, University of Otago

Ex Land & Sea, 16 February 1889.

Swamps remained undrained and even the higher ground turned readily to mud. One disgruntled writer knew where to lay the blame, claiming the Town Board sought to make its 'monument in mud' as they were contriving 'to furnish an inexhaustible quantity'. Mud not only encrusted board members, it coated 'with a hazy envelopment their mental as well as their physical attributes'. Warming to his task, he challenged residents with ways to stop the Town Board's 'mud larks', suggesting they put up photographs of board members so that 'every unfortunate frequenter of the mud-swamped streets' would recognise them and give them 'a thorough bath of mud'. Individuals could also be employed to 'dog their footsteps and aid in sousing them'. Their houses and windows could be covered with mud, bakers could be bribed to put mud in their bread, grocers to mingle it with their coffee. Even their servants could be 'got at' to put mud in their beds. 'The citizens have nothing left for it, but to try an active course of reprisals' (*Otago Daily Times*, 28 May 1862: 4).

While the frustration with the all-pervading mud was clear, it was also perceived as a danger to public health. One visitor to Dunedin recorded that in the town centre there was 'a large expanse of putrid mud, mixed with organic matter, animal and vegetable remains, [and] a constant supply of dead fish from the fish-market close by'. The whole decaying, decomposing mass was, he said, 'daily moistened by the sewerage of the town, just sufficient to keep up the more active fermentation of the putrescent mud'. He believed that the warmth of the summer days would 'profusely generate the humid gases of a stifling atmosphere' creating a 'prolific source of malaria and death' (Wekey, 1862: 69–70). In this description the meaning of dirt and its potential danger is clear. Animal and vegetable remains, including dead fish, join with sewage to form a putrid soup, in which firm ground is changed into fermenting mud. To that writer and no doubt to the citizens of Dunedin, the danger of this putrescent mud lay in the miasma it produced. Inhaling these miasma could result in sickness or death.

Swampy land, besides being useless for building, was dangerous because it held stagnant water and created miasma. It also defied the accepted category of 'ground', being at best a boggy mass in the marginal space between the subsoil and surface water and between the hard, higher ground and the water of the harbour. Corbin, in his study of odour and the French social imagination (1994), proposes that the area which most represented stagnation and accumulation was the marsh and, as stagnation was feared, a whole cosmology evolved around it. Boundaries were blurred between the subsoil, the foetid turf that covered it, and the aquatic mass. The whole vile-smelling ooze was to be avoided and wherever drainage was impossible it was important to cover any slime and prevent the escape of effluvia. Corbin extends the concept of the marsh to the metaphor of the 'city-swamp', where the whole urban environment can be viewed as one boggy mass, filled with stagnating rubbish and giving off foul stenches and miasma. In similar vein, Dunedin can be seen as an unkempt city-body, filthy, foul-smelling and incontinent.

Challenging Containment

The capacity of the city-body to hold its boundaries was most seriously challenged by the sudden influx of people with the goldrush. The population of Dunedin doubled in the first six months following the discovery and trebled again in the ensuing three years, climbing from around 2,500 to 15,790 (McDonald, 1965: 51). Wherever vacant land existed on the hillsides, tents and shanties appeared. The Otago Provincial Council attempted to bring a measure of order to the situation by clearing an area in the cemetery grounds, marking it off in tent sites, but most of the newcomers erected tents wherever they liked. Sanitary conditions were appalling throughout the town. An open drain ran from the small hospital in the Octagon reserve to the flat land below. People complained of its offensive stench and of the 'frequently insufferable' smell still coming from the filthy creek by Rattray Street. A newspaper leader claimed:

> At present, many of the back streets and lanes are in such a crowded state of filth that it becomes a wonder how human beings can live from day to day in such localities. The pestilential odours which arise from the stagnant water and decomposing vegetable and animal matter speedily convince one that for stinks, Dunedin can vie with any City in the Colony.
>
> (*Daily Telegraph*, 8 December 1863: 4)

Hocken Library, University of Otago

Princes Street, Dunedin, 1860.

This was not a reputation to applaud. The writer believed that these pestilential odours created an 'atmospheric pollution' which caused disease, pointing out that a number of fever cases had already broken out. In some parts of the town, he said, a 'peculiar pallor effaces the natural bloom of childhood and marks even the face of manhood'. He called for a Sanitary Commission which he felt sure would 'reveal a tale of horrors of which few have the faintest idea and which is positively a disgrace to a civilised community'. In addition to employing 'scavengers' to keep the streets clean, he suggested a Health Officer should be appointed as well as 'intelligent and inflexible sanitary police officers' to enforce sanitary laws. And above all, the citizens should be roused to compel the authorities to adopt the measures immediately.

The call for a Sanitary Commission was heeded and its report in 1865 gave a graphic account of the shocking sanitary conditions of the town. The evidence of James Nimon, Inspector of Nuisances, told of the 'almost utter disregard to cleanliness' exhibited by residents throughout the town, both inside and outside their dwellings.

> The smell in the more populous neighbourhoods is most offensive, owing to the kitchen refuse, night soil and other filth being cast about indiscriminately, where it will remain for months, unless the Inspector happens to visit the locality ...
>
> (*Report of the Dunedin Sanitary Commission*, 1865: 28)

Sub-leasing properties for shanties or tents had created overcrowded areas, without any provision for privies or ventilation between tents or houses. Open drains flowed through the most populated parts of the town, giving off an indescribable stench. The city-body had challenged its containment and the gravest danger lay in its overflowing channels of excrement.

Sewage – the Embodiment of Dirt

Dirt's most compelling form was human excrement. Beside the mud and open drains of the town, there was a clearly recognised danger from miasma arising from sewage. James Nimon reported to the Sanitary Commission that most privies in the town were badly constructed, being merely holes dug in the soft earth, which filled with water and overflowed in wet weather. Liquid soaked into the ground from these cesspools, polluting the wells close by. Into the northern swamp drained privies, cesspools and other filth, from both private and public drains. Near to this swamp was a 'sea of filth and nightsoil', where hundreds obtained their domestic water from a spring.

In considering the overall problems of sanitation, the Commission received three reports from engineers, outlining possible sewerage schemes and suggesting outfalls into the harbour or the ocean beaches. Although the Commission made recommendations, no concerted action was taken and the dreadful sanitary conditions remained. Ten years later, in January 1875, Dr George Cole, who was Medical Officer of Health for a few months, reported to the City Council that although the construction of sewers had begun, 'our sewage poisons our houses, our streets are filthy and our flat in many parts is a saturated mass of pollution' (1875: 9). Some closets (privies) emptied into the Rattray Street sewer, some into the streets, while others were emptied by nightmen who took the nightsoil to the manure depot. In November the same year Dr J. Gillies reported that many people were still burying their nightsoil in cesspits dug in the ground, shared by up to 20 households. Closets were in an indescribable state of filth. He recommended that every householder should adopt the pail system, with contents emptied into airtight tank carts. Ultimately he hoped that Dunedin sewage would become 'an important article in the agricultural market' (1875: 4–5). People continued their haphazard methods of sewage disposal, however, and it was 1912 before the Council ended all nightsoil removal and considered a reasonably satisfactory sewerage system to be complete.

Whatever schemes were suggested or adopted, the prime concern was to find an effective system for the removal of sewage. Excrement was the very *embodiment* of dirt, representing the city-body's filth in its most marginal form, physically and symbolically. As such, it presented a physical danger, through the noxious miasma it created and a symbolic danger, as the form of filth which most problematically crossed the city-body's internal boundaries from the private domestic sphere to the public arena. It was dealt with through the deliberate use of marginal space.

Hocken Library, University of Otago

Princes Street, Dunedin, 1862 or 1863.

Using Marginal Space

Marginal spaces existed wherever the outer boundaries of the town could be defined. Valley outlets on the perimeters to the north and south, the swampy land at the northern end of the town site and the flat marshy land at the southern end running to the sand-dunes on Ocean Beach, were all marginal spaces. The harbour itself presented a further marginal space between the site of Dunedin and the open sea. Most of these spaces were considered and subsequently used for disposal of sewage. When the Sanitary Commission reported in 1865, it included the suggestions received from the engineers. J.M. Balfour, Marine Engineer for the Province of Otago, recommended an outfall on Ocean Beach as the ideal, but recognised that the City Council's preferred choice might be the harbour. C.R. Swyer, Provincial Engineer, suggested reservoirs near Ocean Beach where sewage could be treated and the material sold to farmers. The final report, in florid prose from J. Millar, the Dunedin City Engineer, recommended Lawyers Head near Ocean Beach as the outfall site. The Commission ruled out Ocean Beach and Lawyers Head on the grounds of cost and risk of pollution, recommending instead an underground sewerage system which would empty into the harbour, with water-tight cesspools for those houses not connected to the system (1865).

In the end the Dunedin sewerage system did discharge effluent at Lawyers Head, but in the intervening years sewage emptied into the harbour, to the dissatisfaction of the Harbour Board. At one time the City Council experimented unsuccessfully with a scheme for conveying the sewage by punt to the lower harbour for tidal removal, an experiment which drew complaints from local residents and the nearby Port Chalmers Borough Council. Many household privies, pans, pails and cesspits were emptied by nightmen who delivered the nightsoil to manure depots set up on the margins of town – in the valley outlets to the north and south and on the flat land near the southern end of town. The depots received a variety of waste but even here there were specific categories of dirt. Depot attendants refused to accept waste which did not fit their definition, turning away cartloads of fish offal and the occasional dead horse.

While there were pragmatic reasons for depositing sewage into the harbour or ocean, or at manure depots on the urban outskirts, these sites are significant in that they were at the physical margins of the town and represent the symbolic margins of the metaphoric city-body. In particular, as a site for dumping excrement the harbour was the large but ineffective bowel of the city-body, failing to excrete the sewage channelled into it. Yet at these physical boundaries of valley outlets, swamps and harbour, society was defining the margins of both its physical and metaphoric body and it was here, in these marginal spaces, that the settlement chose to deposit its embodied dirt.

Marginal Occupations

The workers associated with the removal of waste were themselves marginal. By dealing with refuse and excrement they aligned themselves with filth and in so doing removed themselves from the category of socially acceptable work. The nightmen who removed the contents of pails and cesspits, the 'scavengers' who cleaned the streets of their filth, rubbish and dung and the men who tended the manure depots, all occupied marginal positions. Corbin describes similar workers in Paris who exuded an animal smell derived from their association with filth. These 'comrades in stench' were the prostitutes, ragpickers, dung-men, sewermen, knackers, drain cleaners, refuse dump workers and dredging gangs, who worked with slime, rubbish, excrement and sex (1994: 144–146). For the men in Dunedin, the nightmen, scavengers and manure depot attendants, public perception ensured that they too remained firmly in the realm of filth. Their activities were to be hidden from the public nose and gaze and were controlled by stringent regulation.

These men worked within the town's marginal spaces – the manure depots on the outskirts, the public spaces of the streets and in the boundary between domestic and public spheres. This was a blurred boundary because nightmen necessarily had to work at the fringes of domestic space, entering private premises to remove pails or empty the innumerable cesspits.

Crossing Boundaries

Another to cross the boundary between public and private spheres was the Inspector of Nuisances who, on behalf of the City Council, followed stench and excrement into the home to sniff out foul-smelling drains and overflowing cesspools. The prevailing definition of filth therefore allowed public access across the internal boundaries of the city-body into the private domestic realm. Corbin describes a reorientation within public health discourse in nineteenth century France, away from the public sphere toward private space. Olfactory observers no longer limited their vigilance to the public stench of swamps, stagnant water and sewage, but shifted their attention to the human swamp, now monitoring stench in the dwellings of the poor (1994: 142–143). In Dunedin, the Inspector of Nuisances' interest in the private realm of households was apparent from the beginning of his inspectorship in 1861. By the time of his evidence to the Sanitary Commission in September 1864, he had served 870 notices, summoned 962 people and had initiated prosecutions resulting in fines of £595.5.0. Although initially his attention focused on the earlier-populated southern part of Dunedin, as settlement extended north the number of nuisances increased fourfold. Heavy fines had little or no effect on owners or occupiers in impelling them to keep their premises clean. Many 'deposited their nuisances' in the public street (*Report of the Dunedin Sanitary Commission,* 1865: 28).

Other sanitarians were also interested in the domestic sphere. In his report to the City Council in 1875, 10 years after the Sanitary Commission's report, Dr George Cole claimed that many houses in the city were still 'positively not fit for animals to dwell in' and that, despite good drainage and sewerage, there would always be a high mortality rate unless this was remedied. He advised that no one should be permitted to build a house with fewer than three rooms and that ventilation 'according to sanitary rules' should be required for domestic and public buildings (1875: 4). In the same year Dr J. Gillies reported to the Council that during the previous year four cases of typhoid fever were admitted to the hospital from one residence known as the Barracks, where 'a small porch *at the back of the kitchen* ... [had] been made a receptacle for nightsoil' (1875: 6). Not only were sanitarians entering the home, they were defining the organisation of space on medical grounds.

Sanitarian surveillance therefore extended across the internal city-body boundaries between public and private. Just as it brought an enquiring public gaze into the private sphere of the home, it also attended to private needs in the public arena. From 1862 the Town Board took responsibility for providing public toilets. The first urinal was built, perhaps not surprisingly, near the Town Board office, with others in the marginal public spaces of the harbour jetty and Octagon reserve. The need for public toilets for women did not become the subject of public discussion until 1898.

Another public area to which sanitarian surveillance extended was the school. Corbin links private and public spheres through the medium of the school by contending that enforcement of discipline in defaecation in schools prepared for the diffusion of these attitudes into the private domain (1994: 174). While the inculcation of personal habits in this way was not a focus in Dunedin schools, school inspectors were certainly interested in seeing that standards of cleanliness were maintained in school toilets and in ensuring that they were not used for graffiti, as they considered them cesspits of physical and moral danger. Schools did provide a convenient arena for public health information. By educating the children, information could be disseminated to the homes, expanding sanitarian influence beyond the public domain.

Dirt, Marginality and the City-body

Dirt pervaded all aspects of life, both public and private and necessitated vigilance on the part of public authorities as well as private individuals. As an intrinsic component of nature and an inevitable product of human activity, dirt needed to be either contained or removed as its presence signified risk, recklessness and a disregard for health. More than this, dirt betrayed human efforts to control and impose a structured order on nature. Settlers hoping to leave behind the dirt of cities found that filth followed them into the new settlement, overwhelming them in its abundance. Sanitarian interest in dirt permeated the public domain of swamps, streets, toilets, public buildings, schools and sewers, as well as the private domain of the home. While interested in clean water and food, fresh air and ventilation, sanitarians' main concern was the removal of society's filth. Excrement was embodied dirt, providing the link between private and public portions of the city-body. It crossed the internal and external body boundaries and represented filth in its most marginal form.

As with any metaphor, the value of the 'city-body' with its 'body boundaries' lies in both its fit with a given situation and its discontinuities. The settlement of Dunedin can indeed be viewed as a natural system, an organic body, yet not one in which separate organs functioned smoothly together for the benefit of the whole organism. Nor did its growth follow a measured plan or developmental sequence. Growth and decay emerged together as the stench of rotting waste clung to fine new buildings and open spaces. And any orderly development that had been envisaged by settlers was suddenly and dramatically overtaken by the goldrush, throwing the city-body off balance in a way that could not be rectified through homeostatic adjustment. Becoming incontinent, it spilt its waste wherever it could. Some measures for cleaning and containing waste were effective in the short term, such as manure depots, or protracted in developing, such as the underground sewerage system, but the harbour remained an ineffective bowel, clogged and foetid. If the settlement of Dunedin is to be viewed as a body at all, it is perhaps best viewed as an unkempt and unwieldy body, filthy, soggy, rotting and incontinent.

No one metaphor can be expected to explain all the subtleties of a complex system, nor should it be slavishly followed. Metaphors are useful in so far as they allow a play of ideas. Some aspects remain as yet unexplored, such as the dimension of gender. It may be fruitful to examine the separation of public sanitary work, which had high visibility and was the province of men and the expectation that dirt within the more hidden, private, domestic sphere was the responsibility of women. This exploration might also be informed by the writing of Kristeva who, drawing on Douglas's work, describes the abject body, filth and defilement and their relationship with the female body (1982). Other aspects of the metaphor are most useful when expanded, such as the idea of dirt, or when mixed with other notions, like marginality. Yet as a device for exploring ideas and taking notice of particular aspects of a situation, the metaphor of the body, with its body boundaries, does indeed act as an effective beginning for the exploration of aspects of public health in nineteenth century Dunedin.

References

Cole, G.W. (1875), *Preliminary Sanitary Report of the City of Dunedin*, Dunedin: Dunedin City Council.

Corbin, A. (1994), *The Foul and the Fragrant: Odour and the French Social Imagination*, London: Picador.

Daily Telegraph.

Davison, G. (1983), 'The City as a Natural System: Theories of Urban Development in Early Nineteenth Century Britain', in Fraser, D. and Sutcliffe, A. (eds), *The Pursuit of Urban History*, London: Edward Arnold.

Douglas, M. (1973), *Natural Symbols: Explorations in Cosmology*, London: Barrie and Jenkins.

Douglas, M. (1975), *Implicit Meanings: Essays in Anthropology*, London: Routledge and Kegan Paul.

Dunedin Sanitary Commission (1865), *Report of the Dunedin Sanitary Commission*, Dunedin: Otago Provincial Council.

Foucault, M. (1989), *The Birth of the Clinic: An Archaeology of Medical Perception*, London: Routledge.

Fulton, R.V. (1922), *Medical Practice in Otago and Southland in the Early Days*, Dunedin: *Otago Daily Times* and Witness Newspapers Co.

Gillies, J. (1875), *Report of Removal of Sewage*, Dunedin: Dunedin City Corporation.

Kristeva, J. (1982), *Powers of Horror: An Essay on Abjection*, trans. L.S. Roudiez, New York: Columbia University Press.

McDonald, K.C. (1965), *City of Dunedin: A Century of Civic Enterprise*, Dunedin: Dunedin City Corporation.

Otago Daily Times.

Otago News.

Wekey, S. (1862), *Dunedin As It Is*, Melbourne (no publisher).

The Body of the Land – the Bodies of the People

Gender in Tikopia

Judith Macdonald

> ... to participate bodily in everyday practical tasks was a creative technique which often helped me grasp the *sense* of an activity by using my body as others did. This technique also helped me break my habit of seeking truth at the level of disembodied concepts and decontextualised sayings. To recognise the embodiedness of our being-in-the-world is to discover a common ground where self and other are one, for by using one's body in the same way as others in the same environment, one finds oneself informed by an understanding which may then be interpreted according to one's own custom or bent, yet which remains grounded in a field of practical activity and thereby remains consonant with the experience of those among whom one has lived.
>
> (Jackson, 1989: 135, emphasis in the original)

Through long-term involvement by anthropologists with a field community, learning occurs in part by mimesis, an active immersion in another way of life. Participant observation – taking part in daily activities as well as seeing the patterns of movement – helps the fieldworker approach an understanding of the structure and texture of the everyday, the cyclical rituals and the unexpected. Acquisition of the language of the people with whom they live adds – to the experience of seeing and doing – a commentary from those who teach, guide and inform them.

In 1980 I spent most of the year on Tikopia, an isolated Polynesian outlier in the Solomon Islands. On this island the population of 1,200 lived almost entirely by subsistence, catching fish and gardening. The houses were made from thatch, the clothes from barkcloth although woven fabric was increasingly being used as workers returned from copra plantations elsewhere in the Solomon Islands. There was no money economy and virtually nothing on which to spend money; cash earned elsewhere was usually converted into goods to bring back to the island – cloth, fishing line and hooks, kerosene and water containers. The only contact with the outside world was a ship which came every two months, and there were no resident Europeans. Although there was a knowledge of the wider world of the Solomon Islands, especially among the men, it rarely impinged on life in Tikopia and ties of kinship articulated all relationships between humans, with the land and with the spirit world.

My central interest was in the lives of the women so my days were largely spent working and talking with the women who taught me to act like a woman of Tikopia. The women of the family to which the premier chief had assigned me were my first teachers but later many other women helped to educate me in appropriateness. They began with the reorganisation of my dress and deportment. My hair was clipped close to my head as is common for married women and I was given barkcloth for skirts: an old one for working in the oven house or gardening, a newish one for visiting and one dyed with turmeric for dances. A man in my family made me the rope belt used to hold the barkcloth around the body. Tikopia barkcloth is as thick as cardboard, a long piece is wrapped round the middle reaching from midriff to below the knees and the belt tied around the waist holds it together. Immediately my walk changed. The long strides of my trousered Western self were impossible as was easy bending from the waist in a new barkcloth which could chafe and cut. Old working barkcloths were more flexible but the body still felt corseted and it was easier to kneel than bend. Like the other women, I wore nothing else so I could feel the weave of the baskets held against the bare back with carrying straps that compressed the diaphragm and made breathing difficult when bringing vegetables down from the mountain gardens.

I was also taught how to behave in a house, the appropriate disposal of the body in posture and in space. Then I was taught my place in the wider world of our village, other villages, the gardens, the bush and the sea. With all this came the statements about place, the reasons for prohibitions about women entering certain areas as I was corrected for my habit of wandering about the island alone. And so I learned the shape of the environment and my place, which was a Tikopian woman's place, in the landscape. I adopted the slow, legs slightly apart walk of women in a group, the crouching walk to pass a group of men, polite crawling in the house and the deeply respectful posture to be adopted near a chief. The houses, paths, gardens and reef became my habitus, each place requiring a different bodily response.

By the end of my stay I had a picture in my mind of the environment in which people moved in patterns which spoke of gender and status, the relationships between people, with the land and with the spirits. This chapter abstracts the patterns to map the place of women in a landscape that was not only physical but also social and moral. While many descriptions are generalised, they are based on my lived experience, on the educative statements of my mentors and my observations of those with whom I lived and worked. There is also and inevitably, my outsider's view, coloured by Western epistemologies of the body, which categorised the embodiment of Tikopia gender relations in a way in which no Tikopia woman would articulate her place in that world. Their understanding of their place in their society was neutral and taken-for-granted; they did not criticise their assignment nor suggest alternatives. The parameters of the women's landscape were, to them, natural. There was no alternative landscape, except through migration and, in tribal society, being part of the collective is necessary for survival. By contrast, my understanding of my location as one of the woman on the island was defined by the prohibitions, the knowledge that I was excluded from relationships, statuses, places and ceremonies that were open to men and that I was controlled by the men of my family. The idea of control, therefore, colours my interpretation which is also influenced by a knowledge of many choices of landscape for women, not only in my own society but also on other Polynesian islands where life for women is less constrained and confined than in Tikopia. Nonetheless, this is only a question of degree: all societies constrain the bodies of their members and use them to make statements about ethnicity, gender, class and age. These statements can be pragmatic, ideological, political and metaphorical and a body of theory, much of it from assorted feminisms, has been used to interpret the body.

I have taken three theoretical perspectives to provide a framework for discussing the disposition of women in a particular environment: the cultural construction approach of anthropology, a Foucauldian one of control, and one influenced by Mary Douglas which deals with boundaries. Through the body social statements are made: with the body social relationships are demonstrated but the body is not only a text of culture; it is also a direct locus of social control (Bordo, 1990a: 13). Bordo, following Foucault's (1979) distinction between two arenas of the social construction of the body – the 'intelligible body' and the 'useful body' – says:

> The 'intelligible body' includes scientific, philosophic and aesthetic representations of the body, norms of beauty, models of health and so forth. These representations, however, may also be seen as legislating a set of practical rules and regulations (some explicit, some implicit), through which the living body is 'trained, shaped, obeys and responds ...;' becomes, in short, a socially adapted and 'useful body'.
>
> (Bordo, 1990b: 85–86, emphasis in the original)

Bordo also examines the body in the nexus of power relations (1990b) through which female bodies become what Foucault calls 'docile bodies' (1979: 135ff). While Bordo is concerned with Western women and the self-modifying discipline of diet, makeup and dress, her ideas about the body as the locus of social control can equally be applied cross-culturally to reveal the mechanisms which can bind women into their gendered roles.

However, as well as the idea of bodies being controlled and moulded to a society's norms, there is also the symbolic relationship between body corporeal and body politic. As Mary Douglas says, the idea of society is a powerful image and it is an image which has form: it has external boundaries, margins and internal structure. The body is a model which can stand for any bounded system: its boundaries can symbolically represent any boundaries which are threatened or precarious. The powers and dangers which threaten the social structure are 'reproduced small on the human body' (1970: 137–138). Notions about the individual body and the body politic permeate theories which construct the ideological reality within which the subject comes to be gendered, inhabiting a body whose biological and sexual manifestations are coded in specific ways at specific times (Jacobus *et al.*, 1990: 2).

The statements that the Tikopia as a whole make about their identity as Tikopia to non-Tikopia are clearly visible through dress, language and manner. But they make more subtle statements to and about themselves through the construction of metaphorical and social relationships between humans and environment and men and women in Tikopia. I have chosen two aspects of the body in Tikopia through which to examine these relationships. The first is the ways in which the land is conceptualised as a body and the delineation of internal and external boundaries of that body which are also markers of gender specific realms. This perspective begins with the land and moves to the people. The second perspective begins with the Tikopia understanding of the physical body and its social articulation in the gendered landscape. Both perspectives reveal the way in which women's bodies are defined, controlled and made docile but neither is discrete as a dialogue develops between the cultural understandings of the body and the repetition of gendered bodily tasks in a social and geographical landscape.

The Body of the Land

On an island the external physical boundaries are incontrovertible: there is the land, an intermediate zone of lagoon and reef and then the deep sea. The internal boundaries are both natural ones (mountains, rivers) and constructed ones (paths, houses). Tikopia is a small island (a mile by a mile-and-a-half) with a fringing reef around most of its coastline. It has a small lake, swamp lands, gardens and wilder areas on the flank of a steep mountain which is called the 'Head of the Land'. Paths run around most of the island and lead to gardens as well as giving access to some of the bush areas where large crop trees grow. Villages are clustered around the Western and southern coasts, each one made up of 10 to 20 households. It can easily be mapped to show the geographical features and the boundaries of clan lands and family gardens. However, an invisible map lies over the physical one which defines where people, especially women, may and may not go.

This delineation is based on the social structure of Tikopia which is patrilineal and patriarchal. While all the land is officially owned by the four chiefs (who are called collectively 'the body of the land'), families associated with each clan have the use of gardens and house sites on that clan's land. Land and titles are passed through the male line and all positions of authority, religious and political, are held by men. A woman lives in the household of her father until she marries, when she is taken to her husband's house. If she does not marry, she must stay in her father's house and, after his death, her brother's. She never has land or a house of her own on an island which is a male domain.

Women and especially married women, are constrained in this landscape. They may not go in the canoes into the deep sea outside the reef. However, they can operate at the margins of the water, wading in the lagoon to fish with a handnet and fishing in the lake, standing in the shallow water but not from a canoe. On land, they may go between their house and the reef or lake to wash and, with other family members, they go to the family gardens to weed or gather food for the household. Wandering on casual visits is discouraged; women may go to other villages if they are with a family group and there is a specific purpose for their visit. To walk alone around the island, especially for a married woman, is a sign of immorality and that she is on her way to an assignation. Indeed, an illegitimate child is called 'the child of the path', while legitimate children are seen as fruit, growing in clan gardens from the body of women who, on marriage become members of their husband's clan.

In part, the control of married women's movements and their constant chaperonage by kin is intended to ensure legitimacy of offspring.

For the unmarried woman there is greater latitude. All unmarried women from around puberty are called *fafine taka* which translates as 'wandering women'. Their role is discussed more fully below but, as their name suggests, they may move more freely between villages and gardens although the deep sea outside the reef is also forbidden to them. By contrast, men move throughout the whole landscape – sea, clan gardens and village. Only at night is it problematical for a man to go alone into the bush and more isolated gardens where, after darkness, predatory female spirits may attempt to seduce him. If the female spirit becomes pregnant from this encounter she will then cause the man to die so that he will join her in the spirit world and complete her family. Because the island is so small, the Tikopia traditionally controlled their population carefully. It is therefore significant that women who wander alone away from family territory and men who leave the family group in the night are threatened with stigmatised offspring and, in the man's case, supernaturally caused death.

As the land belongs to the patriline, so does the house and, where this domain is in many other societies regarded as the domestic and therefore female sphere, this is not so in Tikopia. The house is normally a large one-roomed thatched structure, rectangular in shape, a line of posts supporting the central ridge pole and dividing the house in two. These poles are called *pou*, a term also used for male children, symbolically the support of the house. The side nearest the sea is the sacred side and important ancestors of the men of the house are buried here. Inside all Tikopia houses are grave mounds covered with mats emphasising the continuity between the living and the dead of each *paito*, a word meaning both house and lineage. The other side is the domestic one where everyday life goes on – related family groups sleep here, food is brought in from the oven house to be eaten, visitors entertained. Men may not only move on both sides of the house but they can also stand up, while women are restricted to the non-sacred area and may not stand up inside the house but must crawl to move around and to enter and leave by their own door. To obtain something hanging in the rafters they must ask a man to lift it down for them.

Even the oven house '*fare umu*', in other societies regarded as purely the domain of women and where the family food is cooked, is under the direction of the men of the household. While women may move more freely here, they still do so in a crouching posture so that their heads are not higher than the heads of the men. Usually men prepare and light the fire to heat the oven stones and both men and women peel and cut the vegetables. Women prepare the leaves for wrapping the food prior to cooking. Meanwhile, young men have climbed the coconut trees to collect fruit and they grate and squeeze the flesh to make the coconut cream for puddings, the most important type of food. Married women never climb the trees and unmarried teenage girls rarely, although as more young men leave the island to find work in other parts of the Solomons, more girls are needed to fill the roles of their absent brothers. Females on the whole do not express the coconut cream. As one woman said to me, this is because women's hands are filthy from looking after children. While women do clean up after babies, this statement has a symbolic component more to do with contact with genitals and waste than a lack of handwashing. It is also compounded by the belief that semen comes from coconut cream; the material that makes the substance of the child must therefore be protected from possible pollution.

The patterns of control and confinement were acted out daily. No women went beyond the boundary of the land into the sea beyond the reef and married women normally moved beyond the house only in the company of kin related through their husbands. These were the social boundaries based on clan affiliation and land ownership. But they were also moral boundaries, the breach of which could lead, at worst, to illegitimate children, at least, to suspicion of a disloyal or disruptive wife. By contrast men could move anywhere and alone, if they chose, without social or moral sanction except for being in the bush at night which could incur danger from spirits. This was in accordance with the metaphorical extension of body/land. The chiefs ('the body of the land') were supported and respected by the people ('the body of the chiefs'); *te kano o paito* 'the body of the house' (meaning the extended kin group) required loyal and co-operative behaviour from the individual members. Disruption of the social environment at any level could cause disruption of the physical environment. When a chief inherits the title from his father or another male relative he may no longer cut his hair and he is protected as much as possible from sickness and injury because it is believed that diminution of or damage to the chiefly body will be echoed in the body of society generally – people will die or become sick. If a chief becomes old and frail, it is believed that the land will not produce well, reflecting this frailty. The chiefs are therefore the embodiment of the strength and wellbeing of land and people.

The health of the extended family equally depends on the social and physical wellbeing of its members. Bad behaviour by one person (for example, a wife's adultery, a son's rudeness) is likely to affect other members of the family. The husband or the father will become ill, a child may die and bad luck at fishing and poor crops is often attributed to quarrelsome behaviour in a household. This perspective of gender relations in Tikopia showed that, on the body of the land, groups of people moved in ways and places making patterns defined largely by gender and that explanations for problems in the natural body are sought in disruption of the social body.

The Bodies of the People

Another perspective begins with the individual and a cultural definition of what makes a man or a woman. This perspective suggests why the Tikopia believe women need to be controlled. Masculinity and femininity are clearly demonstrated in the clothing and presentation of the body – the external markers of difference – and these differences are mediated by the deportment required by the changing social roles of the life cycle. The postures and gestures I learned while living on the island were the ones which enacted gender relationships in every aspect of life and they were underscored by cultural statements (made to me and to every child) about appropriate behaviour.

There are three categories of female: the child '*tamariki*', the unmarried '*fafine taka*' and the married '*fafine avanga*'. The category of unmarried woman begins in adolescence but continues until marriage, or the end of life. It is, like the English title 'Miss', a referent of marital status, not age. For males, the end of childhood is marked by a ceremony for the supercision of the penis in early adolescence. Thereafter they are considered adult and because traditionally many men did not marry, the distinction between married and unmarried men is not so marked as for females.

Small children run about naked but from the time of putting on clothes they are gender-marked: females wear barkcloth skirts and males a barkcloth strip wound between their legs, round their waists and finishing with a flap hanging down in front while their buttocks are largely bare. Men sometimes cover the barkcloth with a fabric wrap. Apart from clothes, the way in which hair is dressed is a primary indicator of gender and marital status but it is also used to make statements about the sort of role its wearer is allowed, or occasionally chooses, to play. The hair length of the women is an index of growing sobriety – the shorter the hair, the greater the degree of control by others. A little girl, obviously controlled by her family, has her hair cut close to her head apart from a longer fringe all around the hairline. The unmarried female, a status allowing the greatest freedom a woman can expect, grows her hair as long as she likes, but a married woman, emerging in public the day after her marriage, appears with a bristling blue scalp where her hair has been clipped as short as possible. Interestingly, an unmarried woman, once she has passed the age when she can expect to find a husband, may well adopt the restrained demeanor of the married woman, abandon the wandering groups of the younger women and cut her hair short to indicate her new gravity and commitment to her brother's household.

By contrast, men grow their hair long from the end of childhood. It was traditionally bleached golden with lime and during dances the men tossed their heads to display great waves of hair. Most adult men nowadays, while no longer bleaching their hair, continue to grow it long and wear it in a bun for work and loose for dances and social occasions. As was noted before, the chiefs must not cut their hair from the time of their accession and hair was regarded as such an important attribute (for men to grow and women to lose) that one of the clan gods took a special interest in the hair of the people. At the death of a close relative, women, and especially those who were unmarried, cut off all their hair and men, with the exception of chiefs, cut some of their hair. This underlines the importance of hair for making a statement about social relationships with others. Cutting or growing hair makes a further statement about one's status as does the ability to choose and control its length.

There is also a gendering in the general deportment of Tikopia. It is accepted that young children are noisy and run around. Adults therefore keep them away from serious matters and distract them from infringing the *tapu* of male heads and from playing too carelessly with male possessions like knives and pipes. Among the unmarried adolescents and young adults of both sexes there is less gender distinction in behaviour apart from the geographical restrictions which apply to all females. They are a noisy swaggering lot, especially out of earshot of their elders. They work hard all day and dance all night – or so it appears. Their banter is full of sexual innuendo and giggling and they move around the island in groups setting up picnics and expeditions to remote parts of the island or distant gardens.

The married men, especially those of high rank, are grave and dignified on the whole. They do not raise their voices often, they walk firmly along the narrow paths of the island and women and children step off the path to let them pass. Only in the presence of chiefs do men adopt humble positions in order not to be either too close to the chief or higher than his head. By contrast, the married women behave modestly and, quietly in the presence of men, which is most of the time. They crawl if they need to pass a man in the house and crouch if they pass too near men sitting outside.

They sit correctly. *Nofo fakalaui* 'sit properly' is the most commonly heard admonition on the island and it applies especially to married women who may only sit in one of two ways: with the legs together and outstretched or *fai fetu,* with the legs together and bent to the side. And they must not lean on the house posts. In groups by themselves their behaviour can be both rowdy and bawdy but the occasions for this are rare, apart from the licensed carnival before the turmeric-making ceremony. All males sit cross-legged (and may lean against house posts) and sometimes the unmarried females sit that way although it is seen as rather tomboyish. Humble body postures (drooping, crawling, bowing the head and sitting in a constrained fashion) characterise the female side of everyday male–female relations, echoing the respectful postures men must show to chiefs in this hierarchical society and suggesting that, as chiefs are to men, so men are to women.

The internal body and its functions are a matter for speculation. Lacking any type of surgery and having no large animals whose butchery provides analogues, the Tikopia can only surmise from the outside the structure of the inside. They believe that in men the food goes into the stomach but that fatty parts of the food (mainly coconut cream) is stored in the buttocks as semen. A young man with lean buttocks is teased as a sexual spendthrift. It is also known that a woman has a place for a baby to grow. In common with many strongly patriarchal societies, the Tikopia believe that babies are made by men from sperm and that the woman simply provides a place in which the child is formed. I heard no explanation about what happens to the menstrual blood during pregnancy.[1] However, there are symbolic statements about the physical pragmatics of genitals and menstruation which reinforce gendered embodiment.

The definition of Tikopia femaleness appears to have at its core the contrast between the sacredness of the man's head (the *tapu* of the head observed in most Polynesian societies) and the 'dirtiness' of the woman, especially her genitals. The concept of female 'pollution' which occurs especially in anthropological writings about certain Melanesian societies has been challenged as a doubly androcentric interpretation – male anthropologists repeating their informants' fears of female power. It has been suggested that, rather than interpreting women's seclusion in menstrual huts outside the village perimeter as banishment of women by men to avoid pollution of the village, the situation could be read as women voluntarily taking themselves away from weaker men and children at a time of great female potency. This could be seen as analogous with the Polynesian concept of *mana*, a mystical power which is capable of damaging others. In Tikopia, however, women share the view of men that their bodies and especially their genitals are 'dirty' *kerekere* and they call menstruation *toto pariki* 'bad blood'. Firth (1967: 91) quotes an insult song, sung by males against females:

> Your *puka* berry, [a small pink berry, i.e. clitoris]
> Conceal it and let it be hidden.
>
> Don't walk with your legs apart
> Conceal that which smells.

The male belief that female sexual organs smell is frequently repeated in the taunting songs. Males also believe that contact with female organs is dirtying or polluting and women must be careful never to step over men. This idea has supernatural justification. One of the origin stories deals with the birth of the brothers who became the clan gods. The one who should have been the youngest was born first by bursting through his mother's head while the others were born in the normal fashion. The First-Born laughed at his brothers and called them dirty-faced because of the mode of their birth through the genital region. While some contact is unavoidable in sexual intercourse, there is no foreplay and ideally the man will achieve erection himself and the woman will then guide his entry so that his hands do not touch her genitals. A man who touches women's genitals is accused of having dirty hands, an accusation that can be woven by the women in their songs against the men.

Menstrual blood is regarded by both men and women as dirty and polluting socially. A menstruating woman is not supposed to sit near men because 'her blood will leap out towards him'. She is also barred from participation in some ceremonies such as the turmeric making and, in an indigenous accretion to the Anglican religion practised on the island since the 1950s, menstruating women are not supposed to attend church because, as one woman told me, the blood of women must not be where the blood of Christ is. However, there is some ambiguity in the attitude to menstrual blood. While women usually refer to it as the 'bad blood', it is also called *fano* which means 'rinsing or cleansing'. A woman who fails ever to menstruate is called *fafine sai* 'the blocked or closed woman' and it is believed that the few women who come into this category will soon die, presumably because they cannot be relieved of this dangerous substance.

Bordo, cited earlier, took Foucault's idea of the 'intelligible body' (how it is understood culturally) which is then trained to become the socially adapted and 'useful body'. In Tikopia, the female body is understood to have dangerous properties, therefore justifying control and exclusion from certain places and statuses for the health of both men and the land. However, it must also be made useful for production and reproduction within these constraints. That is achieved through a strict delineation of gendered social roles which began at birth.

The umbilical cord of the new baby is cut on the handle of a fishing net appropriate to its gender (the male – flying fish net, the female – reef net) and the placenta of the girl child is buried behind the oven house, while that of a boy is buried in the garden. The dried up cord of a boy is also put in the gardens or in the sea, while that of a girl goes into the garden or on the reef, the areas in which they would operate as adults. Traditionally, at birth a baby has its expected tasks recited to it by the mother's brother. Firth (1967: 51) cites one such recital made to a female child:

> Light be your eyes from the fire which is set up
> For the doings of women
> To be adept at the weeding of taro
> And the making of fine mats
> And the plaiting of floor-coverings
> And the handling of the scoop net
> That the fish wherever it goes

May be caught by you
And when you have brought in your hand-net
Run hither and kindle the oven.
May you be adept in bark-cloth beating
The preparation of *masi*
The planting of the turmeric
All the work of females to be grasped.

It was recognised that the newborn child would not understand the exhortation addressed to it but these messages about appropriate female tasks would be reaffirmed throughout the girl's life. Similar rituals were performed for boy babies, telling them that their work would be in the deep sea and in the making of houses and rope. As the child grows, its gender role is constantly reasserted – a child imitating the behaviour of the opposite sex is not considered amusing as it sometimes is in the West and it is immediately corrected. A little boy who picks up a fan and dances has the fan taken from him and is given a stick which is more similar to the dance paddles used by men. Little girls learn to carry bundles on their back with a carrying strap as women do while a little boy must carry his load on a stick over his shoulder as men do. From quite an early age boys are encouraged to join the men, while girls stay with the women's group.

From puberty women are either *avanga* 'married' or *fafine taka* 'unmarried'. They can also be characterised as the fixed and the free. The unmarried women lead a life of relative spatial freedom in that they can go to most places on the island. They also have the bodily freedom to grow their hair, sit cross-legged and indulge in frequent and varied sexual activities. By contrast, their married sisters are the most controlled group of any on the island. This matter of freedom of movement versus fixity was a central lexical issue. The words used of unmarried women deal with mobility: *taafau* 'wandering, visiting, travelling' and *taka*, which in the term, *fafine taka*, translates as 'mobile or wandering women'. The word *taka* 'to move' can be used in *takavare* 'moving about aimlessly' or, in another sense, if nominalised, as *takanga* 'a coming together, a gathering, especially of the young'. The two words *taka* and *takanga* describe the young women perfectly as they roam about the island in groups. However, even that idea of freedom to move has its limitations: the word for madness is *te pae*, a term which means one who has diverged too far from the path.

For the married woman the central word is *nofo* 'to sit or stay'. *Nofo te avanga* is a frequently heard phrase and means 'the married woman sits, is fixed in one place'. It is often coupled with the phrase *poi ki te ara te tangata* 'the man goes on the path'. In Victor Turner's (1967) term, the path can be seen as the liminal zone between defined areas such as the house, the gardens, the sea or the church. For men, the right to go on the path describes their freedom to move through all parts of the island, but the relationship of females to the path is a much more dangerous one. If the domain of females is the house, in as much as any place is their domain, the paths are liminal zones for them, the boundary between home and wider (and wilder) territory of men. For women danger lies on these boundaries (cf. Douglas, 1970: Ch. 7). Women who go on the path alone are thought to be vulnerable to attack by other men. *Fafine taka*, who may be abducted for marriage or become pregnant, thereby precipitating marriage, are free to go into this semi-dangerous zone of male control but they take the risk of being left with a 'child of the path'. The married woman, already taken, should not put herself in this position. This is, of course, referring to the path as a specific area in ideology or symbolic thought. In practice everybody has to move on the paths but a married woman should do it, as discussed earlier, in the company of others, preferably from her husband's family, and for acceptable purposes such as bathing or gardening.

While the young women work hard in family gardens when their labour is required, Tikopia is a fertile island and a few hours a day is sufficient time to plant, weed and harvest as necessary for the season. Thereafter the young men and women may entertain themselves with expeditions and dances. There is, however, a tense note underlying these youthful activities. Because many of the young men have left the island to work in copra plantations on other islands and young women only rarely leave, there is shortage of potential husbands in Tikopia. Only one of every three girls will marry, thereby gaining the only status a woman can achieve, and the competition for a husband is fierce, causing stresses both between the competing girls and between male and female. In part the stress comes from the girls' attempts to get pregnant and thereby force marriage to some boy. The boys are not so interested in marriage because they hope that one day they can leave the island for work elsewhere, something that is done more easily without domestic responsibilities.

Because of the belief that babies are made of sperm and that a repeated build-up of sperm from one man is necessary to make a baby, contraception in Tikopia consists of young men having intercourse with a different girl each night. One encounter cannot, in their belief, make a baby nor can one be made from the sperm of different men. I have heard boys, discussing the coming night's activities, say 'I had her last night. You'd better have her tonight'. The girls, on the other hand, try to build more permanent sexual relationships in the hope of pregnancy. Therefore the amount of premarital sex is considerable and, being based on very different expectations of outcome, stressful. This is best illustrated in the night dances of the young where *tauangutu* 'fighting mouth songs' are sung.

Tauangutu are dance songs performed alternately by groups of girls and groups of boys, the two sexes taking turns to insult one another. These occasions provide an excellent opportunity for personal rivalries and heartbreaks to be aired and more generalised statements of the relations between sexes. Stereotypes of sex roles in New Zealand Pakeha society are developed in part through the simple crudity of the dirty joke – amazing performance and incongruous conjunction – which is counterpointed by accepted wisdom of the 'no man wants a slice of a cut cake' variety. *Tauangutu* embody the same dichotomy between desire and social practice. Boys criticise girls for their promiscuity and rapacious sexual appetite while praising themselves for their strength and desirability. Girls celebrate their virginity in the face of the boys' importunities, their pride in themselves and their discrimination in the choice of partners. They characterise the boys as immature, sexually greedy and boastful. Each group is slanderous about the other and inaccurate about themselves.

The boys may start with a general comment on female behaviour:

> The weaving and circling of the girls who like the *tauangutu*
> Stand up to indicate, call out

Whether you have a sweetheart.
Me, me, me. [And the boys point to themselves.]

While this may seem at first sight a simple little flirting song, in truth it contains several insults. The word for 'weaving and circling' is also used to refer to the swimming motion of eels in the lake. Eels are believed to have had their origin as the chopped up penis of a god and they are associated with male lust and rapacious behaviour. That the girls circle like eels suggests that they are acting like sexual aggressors. This insult is compounded by the word for standing up to indicate or demonstrate, *sava*. Only men make this gesture so that the boys asking the girls to *sava* showed that they were again suggesting that girls practise male behaviour in taking the initiative in courtship.

The girls may then sing back:

The little senior men of Ravenga
Are hurrying to marry.
But growing up has not come to the little senior men
From coming to make a score.

[The term 'to make a score' literally means tallying as in counting fish but it can also be used in the colloquial English sense of 'scoring'. The girls are suggesting that although the boys may be scoring, they are still very immature.]

And from there on it escalates. The boys sing that the girls are wantons who chase them shamelessly, they ask why the girls take such trouble to choose their white nose shell (a sign of virginity) as none of them is entitled to wear it anyway. This is obviously a time-honoured insult as girls no longer wear the nose shell. The girls reply that the boys are like motor boats trying to buzz into any harbour. But, inevitably, some of the boys and girls will marry, the girls' heads will be shaved and their friends will hold a ceremony of farewell for them.

Tikopia Bodies

It appears paradoxical that the body of the married woman should be so closely controlled socially and geographically while the unmarried girls are free to roam the island and indulge in sexual affairs. It is another paradox that, in a society that makes constant statements about women being dangerous and dirty, young men are not deterred in their sexual pursuit. However, the answer to the paradox lies in the cultural understandings which acknowledge the dangers of wild bush-land and uncontrolled femaleness but control both, as much as possible, to make them useful in production and procreation.

Nonetheless, there are areas that remain wild until brought under control and equally there are the unmarried women who will eventually be definitively assigned to a husband's family or will remain in their brother's household. Until this assignment is made, they have a degree of freedom or wildness, while the married women, like gardens, are at the service of the patriline. This is not a simple nature/culture contrast which posits women's bodies as 'natural' and therefore needing to be brought under cultural control. Rather, it is a recognition of the changing location of women in non-equivalent positions in a substantive web of connections (Haraway, 1991: 191–198).

Through the metaphorical extension between body and land, this understanding is enacted daily, every task and movement in the landscape embodying multiplex social, political and spiritual relationships.

Note

1. More egalitarian tribal societies suggest that the blood and muscles of the new child are made from the mother's menstrual blood while the bones and other hard parts are made from sperm.

References

Bordo, Susan (1990a), 'The Body and the Reproduction of Femininity: A Feminist Appropriation of Foucault', in Jaggar, Alison M. and Bordo, Susan R. (eds), *Gender/Body/Knowledge*, New Brunswick: Rutgers University Press.

Bordo, Susan (1990b), 'Reading the Slender Body', in Jacobus, Mary, Keller, Evelyn Fox and Shuttleworth, Sally (eds), *Body Politics*, New York: Routledge.

Douglas, Mary (1970), *Purity and Danger*, Middlesex: Pelican.

Firth, R. (1967), *Tikopia Ritual and Belief*, London: George Allen and Unwin.

Foucault, Michel (1979), *Discipline and Punish*, New York: Vintage.

Haraway, Donna (1991), *Simians, Cyborgs and Women: The Reinvention of Nature*, New York: Routledge.

Jackson, M. (1989), *Paths Toward a Clearing: Radical Empiricism amd Ethnographic Inquiry*, Bloomington: Indiana University Press.

Jacobus, Mary, Keller, Evelyn Fox and Shuttleworth, Sally (eds) (1990), *Body/Politics: Women and the Discourses of Science*, New York: Routledge.

4

JUNGLE FEVER

AIDS and the Peter Mwai Affair

Heather Worth

> We need to locate the origin of a disease, since its source, always distant from ourselves in the fantasyland of our fears, gives us assurance that we are not at fault, that we have been invaded from without, that we have been polluted by some external agent.
>
> (Sander Gilman, *Disease and Representation*, 1988: 262)

Introduction

AIDS is unique to our era. For the HIV virus, whether or not we are in an 'at risk' group, has meant that no human being is free from its effects. HIV highlights the paradoxical intersections between self and other, bodily pollution and purity. It accentuates divisions between homosexuals and heterosexuals, and between First World and Third World, between men and women, and between race and gender. HIV has been used to articulate a host of issues and concerns, to signify and indeed to justify the interests and actions of many different state and private institutions. It has accentuated a need to track and classify individuals in a sexual demography. It has precipitated increased surveillance, governance and 'political regulation of the body around what is encoded as *the* supremely private site of accommodation to discipline: death' (Yingling, 1993: 305).

This chapter discusses the Peter Mwai case. Peter Mwai was a Kenyan musician, convicted in Auckland, New Zealand of a number of offences relating to exposing women to the HIV virus. I will examine the ways that race (and in particular 'Africa') was used throughout its reporting as a signifier of HIV and how gender became a signifier of purity in the instance of women, and pollution in the instance of Peter Mwai himself.[1] The Peter Mwai affair was a signal instance of the tensions involved in socio-cultural construction of the virus in New Zealand. I will argue that the plight of women infected by the virus through their sexual contact with Peter Mwai, became entangled with an overwhelmingly racist response to Mwai himself. This response cannot just be transferred onto the shoulders of prurient media and patriarchal policy-makers. The response on the part of women's groups to Peter Mwai was not without its own xenophobia.

The state's reaction to the whole affair was to threaten to tighten immigration policy and procedures to include mandatory HIV testing for immigrants. Public health officials have warned of the dangers of sex with men from Africa and the necessity of mandatory HIV testing of pregnant women who have had sex with men from Africa, Asia or America. The emphasis on Peter Mwai himself, meant that those most affected (heterosexual women) were reduced to purely victims in the ensuing debates around the case. Through all of the 'panic' the case engendered, AIDS itself has come to be seen as something external to New Zealand, as something 'out of Africa', which has been imported to infect the bodies of New Zealand women. 'African AIDS' served to distance the New Zealand male population from both a sense of risk and a sense of responsibility and New Zealand women became purely the passive vehicle through which HIV could slip into the country.

The realities of life in the 'postmodern' global village where the whole planet is caught up in international travel and migration means that the New Zealand state (which is paradoxically being portrayed by the National government as part of Asia – in which HIV is now endemic) has an impossible task in attempting to contain the epidemic by closing national boundaries. As Cindy Patton argues, 'the thinking of the nation obscures the realities of sexualities on the move' (1991: 45). In reality the only HIV prevention option open to today's state is widespread education, based on enabling the community (and in particular women, who have traditionally been in a lesser position of power) to say 'no' to sex without a condom. What might have been seen to be the most important issue arising from this case, the fact that many New Zealand women are continuing to have unprotected sex, was forgotten in the haste to undertake exclusionary practices. And the perfect opportunity, while the tragic effects of unsafe sex were highly visible, for a wide-scale health promotion campaign targeted at women, was never even on the public health agenda.

The Peter Mwai Case

> It is perhaps not inappropriate to observe that the intense public interest and anxiety that apparently surrounds this case has played no part in my deliberations. Community responses to the uncontrolled spread of incurable diseases have been invariably marked by fear, anger, paranoia and a desire to attribute blame. But of course, none of those emotive factors play any part in the judicial process.
>
> (Smellie, J., *Judgement*, Mwai bail application)

> The question of Mwai's innocence or guilt has to be decided in a court of law and we should not be caught up in the racist debate about whether he brought Aids (sic) to New Zealand or got it here.
>
> (Lewis Scott, *Dominion*, 28 January 1994: 9)

Peter Mwai was arrested and charged in Auckland in October 1993 with a number of offences relating to wilful infection and possible infection of women with the HIV virus. The arrest followed a number of complaints laid by women who had had unprotected sex with him. Rumours had been circulating for a number of months that linked his possible HIV-positive status with his multiple sexual partners.

All but one of the six charges were withdrawn in early 1994. The remaining charge was that Mwai 'wilfully and without lawful justification or excuse caused or produced in a named complainant a sickness or disease, namely Human Immuno-Deficiency Virus (HIV)'. Subsequently, the five complainants laid six more charges. He was charged with criminal nuisance, that is, omitting to discharge a legal duty with regard to the life or health of each of the five complainants (who were his sexual partners) and one of grievous bodily harm, causing HIV in one of the complainants. The Crown case rested on a number of points: Mwai was aware of his HIV-positive status; he understood the implications of this (that unprotected sexual intercourse could transmit the virus); and finally, he neither revealed this to his sexual partners nor used condoms.

While Judge Smellie may have been able to leave more irrational fears cited above at the door of his chambers, it soon became obvious that myths surrounding AIDS and Africa would overlay the court proceedings and even become an integral part of it. While the trial was not run in an overtly racist manner – there were no references by the prosecution to Peter Mwai's colour or ethnicity and everything was done to keep the prosecution case rational – there was no need for his Africanness, his otherness, to even be mentioned; his difference was plainly visually signalled. On trial, Peter Mwai's body was constructed as extraordinary. Fanon has argued that the effects of colonialism turn African men into 'an object in the midst of other objects' (1986: 77). Peter Mwai became such an object; his epidermal beacon of Africanness was overlaid with other signifiers of difference. His impassive posture meant that it was impossible to read his reaction to the case. Such an attitude, as Watney has argued in general about African AIDS, signified 'an Africa constituted in inscrutability, a treacherous domain of lurking death and hidden evil' (Watney, 1989a: 46). His numerous sexual liaisons with New Zealand women and his disregard for their safety signified African bodies as vessels of sexual treachery. This monstrousness of Africans and Africa, as both paradoxically natural and unnatural, was emphasised by the fact that Mwai did not appear to be visibly ill. Even though the Crown's expert witness told of how AIDS is like rust in the car, invisibly eating away under the surface, unlike the stereotypic images of AIDS ravaged bodies, Peter Mwai was mysteriously and uncannily healthy. While the prosecution realised that overt racism was not a necessary tactic, the defence surprisingly conjured up images of 'witch-doctors', 'rattling of bones' and 'mumbo-jumbo', portraying Mwai as an illiterate man from a poor village in Kenya who, because of his background, was unable to understand the consequences of his actions. His counsel, Barry Hart, argued, 'His state of mind was that he was simply not aware of the situation' (*NZ Herald*, Friday 9 December 1994: 10). This, together with Mwai's own testimony of medical help being available only through a medicine man and of vaccination of 2,000 villagers being carried out with one needle, must be seen in relation to what Watney calls:

> ... countless anecdotes concerning dirty hospitals, out-of-date medical equipment, the sharing of syringes and so on. What is

> not at stake here, however, is the fact of poverty or the effects of post-colonial disruption throughout central Africa, but the implication that African AIDS can be explained as a byproduct of African medicine.
>
> (Watney, 1989a: 55)

The defence itself legitimated the fantasy of intrinsic correspondences between environment, character and physical health.

The jury found that there was insufficient evidence to prove 'wilful infection' but Mwai was convicted in December 1994 of all of the remaining six charges and was sentenced to seven years imprisonment. While there may not have been any reasonable doubt in the jury's mind that Mwai was guilty of the charges, the question remains of whether justice was done? It is hard to argue 'yes' given the way in which the law operates. The law is probably the most clumsy instrument by which to ensure the health of the nation. As Ron Paterson has argued:

> In a climate of fear and suspicion it is easy to leap to the conclusion that infected people who put others at risk of infection should be thrown behind bars or at least isolated. Yet it is not at all clear that public health is well served and the HIV epidemic slowed by resort to the criminal law.
>
> (Paterson, 1993: 37)

Mwai and the Media

> As a result of this Mwai chap, who knowingly screwed around New Zealand, deliberately attempting to infect white women with AIDS, we finally get an Immigration Minister, Roger Maxwell, saying very sensibly, that compulsory testing of HIV was on the agenda.
>
> (Jenny Anderson, Radio Pacific Talkback Host)

Much has been written on the international media's response to HIV and AIDS. It has been stated that the conservative climate in the major Western nations in the 1980s framed the discursive construction of the HIV epidemic (see for example Frankenberg, 1989; Freimuth, 1992; Kinsella, 1989; McAllister, 1992; Mackinnon, 1992; Watney, 1987). The discursive construction of the AIDS epidemic centres around (at least) two tendencies: the framing of HIV and AIDS in terms of panic; and the consequent desire to increase surveillance of certain sectors of the population. Both these discourses work through a logic of inclusion and exclusion. Until the Mwai affair, in the eyes of the New Zealand public and media, AIDS was a disease of homosexuals or, in the singular case of Eve van Grafhorst, a tragic result of infected blood. In this instance, however, panic logic fuelled the *méconnaisance* of HIV as something that had originated elsewhere and had spread into a once-safe community via the bodies of New Zealand women. Peter Mwai unleashed the population's anxiety about Africa as a deviant and viral continent and Africans as having an excessive and lethal sexuality and that both could signify the source and cause of AIDS and himself became the harbinger of AIDS into the population in general.

Singer argues (1993: 28) that AIDS creates a kind of panic logic which attempts to find immediate and dramatic response to the epidemic. The use of victims who are easily identifiable and to whom blame can be attached has been a common media effect[2] (see for example Drushel, 1991; Patton, 1991; Watney, 1989a; Grover, 1987). AIDS has been spectacularised (see Watney, 1989b). That is, in the media's encoding of the medical details about HIV and AIDS, socially recognisable images are drawn, on which echo earlier scourges such a leprosy and tuberculosis. Thus plague imagery such as the Grim Reaper (see Alice, 1992) is used to turn AIDS into a spectacle of death. For example, Bennetton's use of a man dying of AIDS was used to shock readers into taking notice of their product and buying it. This spectacularisation of AIDS has particularly applied to media portrayals of homosexuals who have been represented as those who are guilty of spreading the disease and to Africans and Haitians, particularly those who are part of global migration patterns, who are suspect for the reasons I related earlier. For example, in an article in *Vanity Fair*, Alex Shoumatoff argued:

> I thought of how HIV must have become airborne – airplane borne – moving from continent to continent. ... I imagined this archetypal communicable disease travelling along the mutually manipulative interface of the First and Third worlds in countless copulations.
>
> (Shoumatoff, 1988: 105)

This imagery of AIDS being borne, skywards, from the Third to the First World draws attention to what Watney calls 'profound unacknowledged fears concerning the (nonexistent) miasmatic transmission of the virus which are rationalised and displaced across international travel' (1989a: 51).

It could be argued that in this country there has been surprisingly little homophobic reporting around HIV. The media has, on the whole, portrayed those as most at risk and those with the virus (in New Zealand, mainly homosexual men) in a sympathetic light. On the other hand, the intense media interest in Peter Mwai, as an African man with HIV, was particularly critical. From the first report of Mwai's arrest, media interest was high and was quick to capitalise on the 'exotic' nature of the case. The reporting of the case slid into a melodrama where Mwai became the putative *carrier* of the disease, while the women he infected became purely AIDS *victims*.[3]

Subsequent to the trial, most of this country's magazines carried in-depth stories which emphasised Mwai's Africanness. Metaphors of blackness abounded in such accounts: 'the impassive face of Peter Mwai cast a dark shadow over all New Zealand this week' began the *New Zealand Women's Weekly* story; 'Daktari meets Miami Vice' was the description of Mwai in *More Magazine*; 'His skin is a nuggetty black, his hair woven into braids. He's not tall in the normal way of African men, but his body is taut and toned and moves with feline grace' – *Metro Magazine*. Such commentaries, as Watney has argued, redraw AIDS 'in the likeness of older colonial beliefs and values, targeted at the assumed (white) reader' (Watney, 1989a: 46).

The ideological imperative of the media's reporting of the incident was to render Mwai's Africanness (with all its negative connotations) as the source of infection. Thus the media portrayal of the Mwai case did nothing to change the sexual practices of New Zealand heterosexual men and women.

The connection was never made between the general use of condoms and protection from HIV infection. In fact, implicitly, it contrasted New Zealand men's safe semen compared with unsafe African semen.

AIDS and Governmentality

> AIDS is in the process of redrawing the political front lines and the face of politics, the structures of civil society and of the state.
>
> (Jacques Derrida, 1993: 6)

In modernity, government has a dual role: the welfare of the population, but also its surveillance (see Foucault, 1991). We have seen the implementation of devices such as demography to count and track populations and have witnessed the growth of medicalisation and its control by the state. Within modernity public health offers a proliferation of techniques to keep populations alive and healthy, as well as sexualised. Sex and sexuality thus become discursively constituted in science and public policy; an integral part of the administration and management of the body. Foucault describes the modern body as 'saturated with sex'. He argues that, 'the type of power brought to bear on the body and on sex ... acted by multiplication of singular sexualities ... it extended the various forms of sexualities, pursuing them according to lines of indefinite penetration (1990: 47). Foucault describes these changes as forms of bio-power, whereby disciplines such as psychology not only invoke sexual norms but describe and oversee sexual perversions and the growth in public policies that aim to administer and manage the population's sexual behaviour and health.

AIDS has added another dimension to the governance of the population: it has brought death into bio-power's discourse of 'healthy' sexuality (for there is no cure for AIDS) and it has radically changed the way we think about the human body and its capacity for sexual pleasure. Through AIDS, the human body, already encoded by gender and race, becomes a further site of revision and contestation through disciplines and techniques of body management. AIDS, as a problem unique to modernity, is never just a medical problem; it permeates the entire community and precipitates community crisis. The reaction to this crisis is often and differentially displaced and repeated. It is often itself a 'contagious proliferation' of defensive strategies, aimed at combating the perceived threat to the health of the social body. The logic with which such strategies operate tends to be conservative. Foucault, for example (1973), argues that epidemics can only exist with supplementation by state force and mechanisms of surveillance. AIDS may be used to allow socially authoritative discourses of legitimation to inscribe specific regulatory apparatuses of governance on the bodies of certain sectors of the community whereby social policies may be enacted which intervene harshly in the lives of bodies and populations. As Roger Maxwell, Minister of Immigration argued over the Mwai case, '*It's something the state needs to know about*; people with it [HIV] coming into the country – given that it's highly infectious, given that it's a drain on resources' (radio interview, December 1994. Emphasis mine).

Importantly, it is in the effects of the Peter Mwai case that the intersection between the regulatory procedures and techniques of governance and the organisation and administration of race in New Zealand can be seen to be enacted. AIDS tells us, Watney argues (1989b), a great deal about the changing organisational boundaries in the West. AIDS has been harnessed to a hygenicist rationale by which African blackness becomes a transcendental signifier of AIDS as an external threat which must be excluded. While governmentality exists in relation to liberal democracy, it also operates on the Western philosophical assumptions of inclusions and exclusions, sameness and otherness.

Thus the Peter Mwai affair has engendered increasing public calls and sympathy for a range of interventionist and regulatory practices in this country. As Singer has argued:

> Under the logic incited by epidemics, forms of regulatory intervention into the lives of bodies and populations which might in other circumstances, appear excessive, can now appear as justified forms of damage control and prophylactic protectionism.
>
> (Singer, 1993: 30)

Historically, the entry of immigrants into New Zealand has often been opposed on the basis of race, colour, national or ethnic origin, or religion; there has often been a 'concern' with the entry of 'outsiders'. The Mwai case brought a wider call for mandatory HIV testing of immigrants and refugees. Underpinning the case for such testing is a belief that there is an intersection between outsiders and the disease: AIDS becomes an imported problem. As Hawke and Hart have argued, 'the externalisation of blame has led many governments and researchers to look for the source of AIDS outside their own national boundary. Scapegoating easily recognisable foreign nationals gains swift acceptance' (1993: 207). Mandatory testing had already been an issue in a number of countries, so it was not surprising that the Mwai case legitimised it as an issue here.

As early as 1993, when it became public knowledge that Mwai, who had then been arrested, was seeking permanent residency, the former Police Minister, John Banks, began the call for mandatory HIV testing of those who applied for permanent residency. And by the end of 1993, the Minister of Immigration, Roger Maxwell, had taken up Banks's call and was already signalling that mandatory testing of immigrants was on the agenda. After the trial, the possibility of mandatory testing was made part of the government review of the Immigration Department. Interestingly, one of the ways the New Zealand Government has rationalised the debate around HIV and mandatory testing is through the new managerial-speak of cost to the health system. The problem of race has been displaced onto the 'predicament' of scarce resources.

The fears that mandatory testing would be introduced has, due to the outrage expressed in a number of submissions to government by groups such as the Ethnic Council and the New Zealand AIDS Foundation, been averted at this time. After reviewing the regulations, the Minister of Immigration, Roger Maxwell, agreed that 'screening refugees on the basis of their HIV status is of dubious value and would be in contravention of New Zealand's international obligations' (Radio Interview, 2ZB, 20 December, 1995). However, the call for screening did not stop with the aftermath of the Mwai case. After 13 African stowaways claiming refugee status arrived in New Zealand in November 1995, the MP for Tasman, Nick Smith, was quick to allege their possible infection with HIV. On talkback radio he argued:

Smith: It's going to take two or three years for [the stowaways] to exhaust all of the legal avenues, of course all paid for by the long-suffering New Zealand taxpayer with legal aid ... [what] I was concerned about is that the lawyer representing these stowaways has told the immigration service yesterday that they are going to refuse these people having HIV tests. ... It seems a joke, because they are illegal immigrants and already here nothing can be done about it ...

Talkback Host: Isn't that amazing, just amazing? Thirteen African stowaways arrive in this country on a boat, no documents, claim refugee status. ... Because they are refugees or because they are claiming refugee status they also do not need to take HIV testing. They don't need to be tested for AIDS.

(*Talkback Radio Transcript*, 1995: 1–3)

What is particularly pertinent about this incident is that it reiterates the xenophobia of the Peter Mwai case: the refugees are African; they must therefore be deceitful, and they therefore are likely to be infected with HIV. And, most importantly, they are costing the New Zealand taxpayer money.

Women as Other in the AIDS Debate

> AIDS is being used to validate the drawing of new lines of power and popular consent which require the fabrication of new Others, to bolster healthy familial identities.
>
> (Simon Watney, 1987: 60)

The HIV epidemic in New Zealand has been overwhelmingly homosexual and the number of New Zealand women infected with HIV remains small. This has had the effect that women have been made a low priority group in government-funded prevention programmes. As Linda Singer argues:

> In this economy of contagion, women's position has been relatively marginalised and derivative, i.e. a consequence of their sexual contact with men. As long as the sexual epidemic is figured primarily as a problem of contagion, women's position can be overlooked or minimised by a prudential logic of relative risk.
>
> (Singer, 1993: 84)

However, in the two years to 30 December 1994, 27 women were diagnosed HIV positive, a rise from a total of 19 on 1 January 1993 to 46. This sudden rise in the number of women diagnosed as HIV positive – there were no heterosexual men diagnosed HIV positive in the 12 months to 30 September 1993 – is in part specifically due to Peter Mwai.[4]

The Mwai case also illustrated the ways in which women are specifically and differentially positioned in the AIDS debate. As Cindy Patton states, 'conflicting constructions of gender form the basis of the logic through which science, the media, educators and policy-makers operate' (1994: 1). What has happened is that women have been appropriated and culturally constructed in certain ways through the Mwai case and in particular in terms of sexual appropriateness, monogamy and possible racially-inappropriate reproduction. As a consequence of the Mwai case, women's relation to sexuality is being reinvested with notions of purity and pollution. Women are being urged to function as guardians of pure sexual exchange. Derek Keith in a letter to Metro argued, 'each of these women was prepared to have unprotected sexual intercourse, whether some of them thought they "had no choice" or not, with a man they had known only a few hours, after meeting in a bar' (Metro, March, 1995: 4). What Keith found shocking is that none of the women felt their behaviour to be 'exceptional or exceptionable'. In other words, what is necessary is that New Zealand women regain their status as guardian of the nation's morality and sexuality.

The anxiety, the panic logic, that the case produced has become displaced into the calls mentioned above for the regulation of sexual reproduction and the promotion of the family as the supposedly exclusive site of safe sex. While I am not accusing public health figures of the jingoistic reactions of some politicians and talkback hosts, it is, however, somewhat worrying that the Mwai case has begun to have a conservative effect in the public health arena. In an article in *New Zealand Doctor*, headed 'GP Vigilance a must to detect HIV in mothers', Nigel Dickson of the AIDS Epidemiology Group is quoted as saying, 'while there is no place for universal screening, [HIV] tests may be in order if the woman has had sex with an African, or South East Asian, an injecting drug user or a bisexual man' (1995: 5). And at least one sexual health worker has advised selective refusal of certain racially nominated partners as a preferred social response to the virus. Dr Janet Say, of the Auckland Sexual Health Clinic, argued in the *Dominion* (27 October 1993: 12) '[a] review of cases in Auckland shows the major problem in heterosexual spread of the virus is where there has been contact with people from countries in Africa, South East Asia and parts of America where the virus is widespread'.

Because epidemic conditions raise the social stakes attached to unregulated sexual exchanges, contagion becomes a major figure in social relationships and social production and the fears of the contagious nature of those 'just off the plane' connect international travel to an equally deep-seated fear of racial miscegenation. There are possibilities of social policies being mobilised which will cover over such anxieties by arguing that safety of women and babies justifies the imposition of regulatory controls. As Singer argues, '[i]n the name of sexual health all sorts of means to control populations that would ordinarily seem unacceptable now seem justified as lesser evils' (1993: 60).

The sudden increase in numbers of HIV-infected women around the time of the Mwai case could have resulted in their needs and the greater political issues surrounding these, being put on the public health agenda. One of the most important concerns surrounding the case has been the issue of consent. Many of the complainants had raised with Mwai the necessity for condoms, but had been unable to convince him to use them and one complainant and at least one other woman (not a complainant) felt that coercion in general had been part of their sexual experience with Mwai, so that some of the women were unable to negotiate safe sex. Some of the other women, however, did not believe force played a part in their relationship with Peter Mwai. Their failure to negotiate safe sex was they 'trusted him'. One of the complainants stated:

> I reached over to my toilet bag for a condom. I showed it to him. He indicated to me there was no way he would wear it. It was more body language than anything. He said something very fast, like 'No, they hurt'. I felt very embarrassed that I'd even asked. I felt like I'd ruined the moment. I put it away.

While the decision of his sexual partners to discard condoms may appear to be because Mwai was African and therefore more violently insistent on condom-free sex than New Zealand men would have been, sex without a condom is, on the contrary, the norm in New Zealand. Unlike the gay community where condoms are an eroticised part of sexual life, their use amongst heterosexuals is still extremely low. Recent studies suggest that condoms are used on average by less than twenty five per cent of the sexually active heterosexual population (see Chetwynd, 1987; McEwan, Auket and Hills, 1988; Paul, Skegg, Smeijers and Spears, 1988). The most recent (ethnographic) study of condom use by Neale, Kehoe and Kerslake (1994) showed half of the heterosexually active women in the sample had not used condoms and most of those who did, only occasionally used them. New Zealand sex workers, who have high awareness of the dangers of unprotected sex, also seldom use condoms with non-paying partners. Woods' (1993) study shows that most women who are used to negotiating safe sex with paying clients do not negotiate condom use in love relationships. These women suggested that men would think they were unclean if they insisted on condoms, or that they would guess they were sex workers, or that their relationship would be ruined by such a suggestion. They believed that either they were not at risk of infection and that they trusted their partner. But most importantly, Woods shows that the women had less negotiating power outside the work setting: partners were insistent on condom-free sex, women were on their own outside the safety of the massage parlour during these encounters.

The behaviour of women involved in the Peter Mwai case, the sex workers in Woods' study and those of the women in Neale's work are very alike and are, I would argue, similar to the everyday experiences of New Zealand women in negotiating safe sex. One of the outcomes of the Mwai case is that due to the anxieties and fears and anxieties he as an African man generated, very little was made of the fact that thousands of times each year New Zealand men engage in unprotected sex, or reject condoms as being uncomfortable. Women are left with a de facto responsibility to act as the barrier to the circulation of HIV throughout the community. But this state of affairs cannot be blamed just on a media which has emphasised Peter Mwai's otherness. The government has failed to take the potential for multiple female infection seriously. While women are the fastest growing group being infected with HIV, the government has not funded any targeted HIV prevention programmes for women.

There is also a cause for concern about how New Zealand women reacted to the Mwai case. As I stated at the beginning of this chapter, women's reactions also became entangled in the racist response to Peter

Mwai. Evidence of this comes in the poster which was pasted up around Auckland. In the form of a 'wanted' notice, the poster exacerbated the fears of a marauding African and implicitly that every African man is on the sexual prowl. Part of the success of the police in being able to induce more than one complainant was due to these posters which requested Mwai's women partners to contact the Auckland Sexual Health Clinic or the police. But there is also a sense in which some of those women who made complaints to the police at this time later began to feel that they were being coerced into taking punitive action against him, and one women in particular, who had had a reasonably long-term relationship with Mwai, felt that the relationship was degraded for the sake of gaining a conviction. She became hostile in the witness stand to what she saw as a personal tragedy being turned into a lynch mob mentality.

While much of the response from women's groups was to support the women Mwai infected, I would argue that this became inexorably entangled in the general xenophobic reaction to the case. The use of the 'wanted' poster technique as a feminist response to the Mwai case can be seen as itself implicated in Eurocentric imperialism and highlights feminism's historical placement in a post-colonial world. Feminism has, until recently, been conspicuously absent from the *international* debates around HIV and AIDS, as well as from AIDS activism. Recently there has been a flurry of Anglo-American publications (see for example recent books on the subject King, 1992; Patton, 1993; Squire, 1993; Triechler, 1991; Doyal, Naidoo and Wilton, 1991; Richardson, 1987; Faden, Geller and Powers, 1991) which bemoan the marginalised position of (Western) women in relation to discourses around the virus and, in my opinion, often overstate the risk of heterosexual infection via bisexual men from countries where AIDS is not widespread.[5]

In New Zealand very little attention is paid to the global problem of HIV where women are now the majority of those infected. Instead feminist discourse and more general discourse around women and AIDS in New Zealand women has portrayed women as marginal to HIV and the gay community who face the brunt of the epidemic as sexist and exclusive. In a way women leapt on the Mwai opportunity because it was *women* who were infected and in doing so political caution flew out the window. It is almost as if women wanted to see themselves at the centre of what has turned out to be the communicable disease that kills the largest number of New Zealanders under 65 years old.

Conclusion

Politics, ethics and policy-making around HIV and AIDS are practices fraught with the apprehension and worry that the virus itself generates. The Peter Mwai case exacerbated these concerns. For it located the origin of the tragedy which had overtaken a number of newly HIV-infected New Zealand women in one man and, I have argued, in his Africannness. Even though it can be argued that women were complicit in the ethnocentric response to the case, this reaction has, ironically, deleteriously affected women. It cannot be denied that in this instance the needs of those women who were infected with HIV were marginalised. It has overtaken the greater claim that every day women in New Zealand find it impossible to negotiate safe sex and every day men fail to take responsibility for their own and their partners' safety. The response to this tragedy, the public health outcomes of the Peter Mwai case, have not been to increase targeted HIV prevention to women, but rather to engender calls for mandatory HIV testing of immigrants, for testing of pregnant women and for women to limit their partners to certain 'safe' groups. Through the ethnocentric nature of the responses to the Peter Mwai affair it is women who have been ultimately the loser. The effect of these calls has been as Watney has claimed, to increase 'the phantasmatic projection' (1989a: 53) of the West, caught in the binaries of race/gender, white/black, male/female, repulsion/desire. Thus, in my opinion, women's place in the HIV debate will continue to be one solely of misinformation and objectification.

Notes

1. The notions of pollution and purity have been used very powerfully in the work of Mary Douglas (1966). While this chapter does not refer to her work, specifically, it is important here to acknowledge her influence in my writing.
2. The scandal of Rock Hudson's outing as both a person living with AIDS and as a homosexual, in 1985, is a very pertinent example of this. For an outline of this incident see Meyer (1991).
3. In the *New Zealand Herald* and other media Mwai is named as an 'AIDS carrier', however, there is actually no such 'thing'. HIV is the same virus in all those infected. In none of the media accounts is Mwai himself mentioned as an AIDS victim; in contrast the media almost exclusively refers to the women as victims. The 'victimisation' of those with HIV has been roundly criticised by AIDS organisations as a non-political way of dealing with those who are HIV positive.
4. There is no definitive statement of the number of women whom he infected, but there seem to have been approximately seven known cases to date.
5. There seems to be a general overstatement of the risk of such a source of infection, and in fact I believe a myth has grown up around bisexual infection. HIV has two major strains: B and E. Strain B is found almost exclusively in homosexual men and has been found to be very vagina unfriendly. On the other hand Strain E, found in parts of the world where AIDS is widespread (and with which Peter Mwai was infected), is easily transmitted vaginally. This accounts for the lack of infection of women by haemophiliac partners infected through blood transfusion in New Zealand in the early 1980s (Worth, H. work in progress).

References

Alice, L. (1992), 'AIDS Bodies, Panic Sex and Other Fatal Strategies', in *Sites: Journal of Radical Perspectives on Culture*, vol. 24, May, pp. 9–20.

Camus, A. (1960), *The Plague*, London: Penguin.

Chetwynd, J. (1987), *A National Survey of New Zealanders' Knowledge, Attitudes and Behaviour Relating to HIV/AIDS*, Wellington: Department of Health.

Derrida, J. (1993), 'The Rhetoric of Drugs', in *Differences*, vol. 5, Spring, pp. 1–25.

Dickson, N. (1995), 'GP Vigilance a Must to Detect HIV in Mothers', in *New Zealand Doctor*, vol. 31, March, p. 5.

Douglas, M. (1970), *Purity and Danger,* Middlesex: Pelican.

Doyal, L., Naidoo, J. and Wilton, T. (1991), *AIDS: Setting a Feminist Agenda,* London: Taylor and Francis.

Drushel, B. (1991), 'Sensationalism or Sensitivity: Use of Words in Stories on Acquired Immune Deficiency Syndrome (AIDS) by Associated Press Videotext', in *Journal of Homosexuality,* vol. 21, no. 1–2, pp. 47–61.

Faden, R., Geller, G. and Powers, M. (1991), *AIDS, Women and the Next Generation: Towards a Morally Acceptable Public Policy for HIV Testing of Pregnant Women and Newborns,* Oxford: Oxford University Press.

Fanon, F. (1986), *Black Skin, White Masks,* London: Pluto.

Foucault, M. (1973), *The Birth of the Clinic. An Archaeology of Medical Perception,* New York: Random House.

Foucault, M. (1990), *The History of Sexuality: Volume 1, An Introduction,* London: Penguin.

Foucault, M. (1991), 'Governmentality', in Burchell, G., Gordon, C. and Miller, P. (eds), *The Foucault Effect: Studies in Governmentality,* Chicago: University of Chicago Press.

Frankenberg, R. (1989), 'One Epidemic or Three? Cultural, Social and Historical Aspects of the AIDS Pandemic', in Aggleton, P., Hart, G. and Davies, P. (eds), *AIDS: Social Representations, Social Practices,* London: The Falmer Press.

Freimuth, V. (1992), 'Theoretical Foundations of AIDS Media Campaigns', in Edgar, T., Fitzpatrick, M. and Freimuth, V. (eds), *AIDS: A Communication Perspective,* Hillsdale, New York: Lawrence Erlbaum Associates.

Gilman, S.L. (1988), *Disease and Representation. Images of Illness from Madness to AIDS,* Ithaca: Cornell University Press.

Grover, J. (1987), 'AIDS: Keywords', in *October,* vol. 43, Winter, pp. 17–30.

Hawke, S.J. and Hart, G.J. (1993), 'Travel, Migration and HIV', in *AIDS Care,* vol. 5, no. 2, pp. 207–214.

King, K. (1992), 'Local and Global: AIDS Activism and Feminist Theory', in *Camera Obscura,* vol. 28, pp. 79–98.

Kinsella, J. (1989), *Covering the Plague: AIDS and the American Media,* New Brunswick: Rutgers University Press.

McAllister, M. (1992), 'AIDS, Medicalisation and the News Media', in Edgar, T., Fitzpatrick, M. and Freimuth, V. (eds), *AIDS: A Communication Perspective,* Hillsdale, New York: Lawrence Erlbaum Associates.

McEwan, B., Aukett, R. and Hills, M. (1988), 'High School Students' Attitudes Towards and Use of Contraceptives', in *New Zealand Medical Journal,* vol. 23, March, pp. 142–145.

Mackinnon, K. (1992), *Politics of Popular Representation: Reagan, Thatcher, AIDS and the Movies,* London: Associated University Presses.

Meyer, R. (1991), 'Rock Hudson's Body', in Fuss, D. (ed.), *Outside/In: Lesbian Theories, Gay Theories,* New York: Routledge, pp. 259–288.

Neale, J., Kehoe, J. and Kerslake, V. (1994), *Condom Usage in New Zealand: A Pilot Study of Pakeha Knowledge, Attitudes and Experiences,* unpublished research report, Victoria University of Wellington.

Paterson, R. (1993), 'HIV as Crime: Is it an Offence to Infect?', in *NZ Doctor,* vol. 11, November, p. 37.

Patton, C. (1994), *Gendering the AIDS Epidemic,* New York: Routledge.

Patton, C. (1993), 'With Champagne and Roses: Women at Risk from/in

AIDS Discourse', in Squire, C. (ed.), *Women and AIDS: Psychological Perspectives*, New York: Sage.
Patton, C. (1991), *Inventing AIDS,* New York: Routledge.
Paul, C., Skegg, D., Smeijers, J. and Spears, G. (1988), 'Contraceptive Practice in New Zealand', in *New Zealand Medical Journal,* 14 December, vol. 101, pp. 809–813.
Richardson, D. (1987), *Women and the AIDS Crisis,* London: Pandora Press.
Shoumatoff, A. (1988), 'In Search of the Source of AIDS', in *Vanity Fair,* vol. 51, no. 7, July, pp. 105.
Singer, L. (1993), *Erotic Welfare: Sexual Theory and Politics in the Age of Epidemic,* New York: Routledge.
Squire, C. (1993), 'Introduction', in Squire, C. (ed.), *Women and AIDS: Psychological Perspectives*, New York: Sage.
Treichler, P. (1991), 'How to have Theory in an Epidemic: The Evolution of AIDS Treatment Activism', in Penley, C. and Ross, A. (eds), *Technoculture,* Minneapolis: University of Minnesota Press.
Watney, S. (1987), *Policing Desire: Pornography, AIDS and the Media,* Minneapolis: University of Minnesota.
Watney, S. (1989a), 'Missionary Positions: AIDS, "Africa" and Race', in *Critical Quarterly,* vol. 31, no. 3, pp. 45–62.
Watney, S. (1989b), 'The Spectacle of AIDS', in Crimp, D. (ed.), *AIDS: Cultural Analysis, Cultural Activism*, Cambridge, MA.: MIT Press.
Woods, K. (1993), 'You have Sex with a Condom. You're Making Love Without: Condom Use by Women Parlour Workers in and out of Work', unpublished paper, Department of Sociology, University of Auckland.
Yingling, T. (1993), 'AIDS in America: Postmodern Governance, Identity and Experience', in Fuss, D. (ed.), *Inside/Out: Lesbian Theories, Gay Theories,* New York: Routledge.

Part 2

The Politics of Bodily Representation

Listening to Girls' 'Voice' and Silence

The Problematics of the Menarcheal Body

Kathryn Matthews Lovering

The menarcheal body of girls has been an ever-present object in the production of knowledge about womanhood since the formation of academic disciplines. For the founder of adolescent psychology, Stanley Hall, it was the 'critical experience' marking 'the great and ineradicable difference between the physical and psychic life of woman and man' (1904: 503–504). For the influential writer on the psychology of girlhood, Helene Deutsch, it was a 'trauma' bringing 'anger, shame, depression, feelings of inferiority and guilt' (1946: 479). More currently it is the 'sign and symbol of womanhood' (Logan, 1980: 286) and 'the signal event unique to female puberty' (McGrory, 1990: 265) which is 'a central aspect of body politics' (Lee, 1994: 343). The significance of the onset of menstruation, menarche, has been spoken and debated interminably since the Victorian era (Lovering, 1994). Yet somehow the girls' voice has been absent because it is claimed that for girls the 'shame that went hand in hand with ignorance' has made 'the bleeding ... beyond common language' (Apter, 1990: 33).

This paper is about the production of this shame and bleeding beyond common language. I discuss how girls talk about menstruation, what they can and cannot say about it and what discourses of menstruation are available to them to explain and understand their menstrual cycle. Although my research on the inscription of adolescents into the discourses of menstruation (Lovering, 1994) has included an historical analysis as well as discussion groups and questionnaires with boys and girls, here I deal primarily with the material from interviews I did with girls only.[1]

Background: Menarcheal Research and Questions

When I started my research into girls and menstruation, I began with questions about how young adolescent girls' self-concept might be affected by menarche and what menstruation might mean to them. My initial approach to these questions was shaped by 'traditional' academic psychology (Rose, 1985; Henriques *et al.*, 1984). Here, claims to truth are based on highly formalised methods of data collection and analysis and in the case of research into adolescents and menstruation, on questionnaires and controlled studies done almost exclusively with girls (Grief and Ulman, 1982). Although references are made to social and cultural influences on girls' attitudes and knowledge, the methodology employed and its underlying theory tend to produce results that are assumed unproblematically to come from inside the individual and to be based on a transparent, nature-given body.

Much of the current research into menarche has been part of a constructive feminist project to address issues of importance to women and we have gained important knowledge from this research. From numerous studies, we learn that adolescents in many Western cultures have primarily negative views of menstruation and lack accurate information (Amann-Gainotti, 1986; Lei *et al.*, 1987; Lovering, 1994; Ruble and Brooks-Gunn, 1982; Scott *et al.*, 1989). There is evidence that a lack of preparation for and/or negative expectations of menstruation can result in girls having negative reactions to menarche (Rierdan *et al.*, 1989) and an increase in negative menstrual symptoms (Skandhan *et al.*, 1988). However, in contrast to both the earlier psychoanalytic formulation of menarche as 'traumatic' (Deutsch, 1946) and the recent social science constructions of menarche as a central '(hetero)sexualising' event (Lee, 1994), studies done by women psychologists with adolescent girls have shown menarche not to be necessarily traumatic (Rierdan *et al.*, 1989, Ruble and Brooks-Gunn, 1982), nor does it appear related to girls' self-esteem or self-image (McGrory, 1990; Weiden, 1984).

As for the research into the social aspects of adolescents' attitudes to and knowledge about menstruation, it is very limited, with little research into family, women's or educational practices. And I would argue that even the research done with girls and women is bedevilled by a number of methodological and theoretical problems including highly selective samples, retrospective studies and unexamined assumptions so that this research, even when feminist, constitutes the onset of menstruation and subsequently the female body as uniquely and essentially problematic; women-only processes and women-only problems.

Even as I started my research with adolescent girls, I had doubts about the impact of menarche in itself on girls' self-concept or identity because of my own personal experience. My first period was so unremarkable and unremarked that neither my mother nor I can remember it; I will return later to this forgetting of what is theorised to be such a memorable and significant event (Grief and Ulman, 1982; Scott *et al.*, 1989). And my research with young British adolescents (Lovering, 1994) supports this doubt, as does previous research: menarche has been found to have no significant affects on girls' self-concept, body image or self-esteem.

These results presented me with a contradiction. Like other women researchers, I knew from my own personal experience how problematic and regulating menstruation can be in our lives and yet research with young adolescent girls into the effects of menarche on their attitudes and self-concept has yielded few significant results. In addition neither questionnaires nor psychometric tests can adequately address the question of the meaning of menarche or menstruation for girls. In order to explore this question and to learn more about how menarche might affect girls, I decided to use a more open-ended method, with in-depth interviews with girls pre- and post-menarche using what Corrine Squire calls a 'woman-centred' approach (1989).

Contextual Analysis: Listening to Girls' Silence

To begin this discussion it is important to stress that it was never intended that the girls in this study would constitute a representative sample: they were 46 girls from South West Britain who were predominately white European and drawn from across the academic ability range and social classes. In terms of their menarcheal status, at the first interview, 40 girls were premenarcheal and six girls were postmenarcheal. At the second interview, 21 girls were premenarcheal and 25 girls were postmenarcheal. For the interviews I used a semi-structured interview schedule because I wanted the interviews to cover the same topics more or less systematically. The topics covered were self-concept and self-regard, possible selves, relationships, menstruation, feelings about growing up and significant life events. In conducting the interviews I tried as far as possible to create an atmosphere in which the girls could 'elaborate on their views in a relatively naturalistic conversational exchange' (Potter and Wetherell, 1987: 165).

I began my analysis of the interviews using what has been referred to as a 'contextual analysis' (Belenky *et al.*, 1986: 16). This approach was adopted by feminist psychologists working on women's 'Self, Voice and Mind' (Belenky *et al.*, 1986) and was based in part on the work of Carol Gilligan (1982). When adopting an analysis of interviews with women only in which 'women speak for themselves' (Belenky *et al.*, 1986: ix), feminist academics are applying the political principle 'that women's experience can provide a direct route to women's consciousness or identity' (Hollway, 1989: 41). And my initial analysis of the interviews was informative about girls' knowledge and expectations of menstruation, for, as Wendy Hollway states, this approach achieves 'a reasonably faithful reproduction of whatever assumptions people use to interpret their own experience in the research relationship' (1989: 40).

However, an inductive contextual analysis continually proved to be problematic for me, in part because it is based on the idealist assumption that 'the knowledge is there, based on experience and can be represented in an account' (Hollway, 1989: 41). In this case the analysis demonstrated that these young adolescent girls had limited knowledge about menstruation and of course that premenarcheal girls had no direct or inner personal experience of menstruation on which to base an account. Nor could the majority of girls talk openly enough about menstruation to present direct accounts to an adult.

In addition, an inductive contextual analysis tends to treat the account of the participants as producing facts which are the 'truth' about the individuals' inner thoughts, feelings or experiences. These revealed 'truths' are not conceptualised as accounts produced by discursive practices and mediated by social and power relations. The effects of the gender of the participants and researcher or the power of males over females or adults over children do not form part of the analysis. Rather, a contextual analysis 'reproduces (and legitimates through science) whatever discourses research participants use to position themselves at the time' (Hollway, 1989: 40).

Using this approach presented me with two problems. First, I was in danger of reproducing the normative social science discourse about adolescents' *failure* to learn the correct biology of reproduction and positive attitudes to menstruation: what I call the menstrual misconceptions and miseries discourse. But second and even more problematic, I found myself unable to account for girls' silence about menstruation or adolescents' construction of the female body as more problematic than the male body. Rather than going beyond attitudes to menstruation to the meanings of menarche (see Potter and Wetherell, 1987, for a detailed critique of attitude theory), my contextual analysis constituted the body and menstruation as an ungendered scientific object about which adolescents have feelings and beliefs rather than as the female body whose difference and femininity is constructed in the social world by power relations. To explain adolescents' lack of knowledge, embarrassment or unwillingness to talk about menstruation and what menarche might mean for girls requires more than an inductive contextual analysis which codes and categorises subjects' accounts as if they are an 'expression of the real person' (Hollway, 1989: 42).

These difficulties made me realise that I had to use a different theory and analysis. Traditional psychological theory even with the addition of woman-centred descriptive interviewing was not going to give my research the insights into girls' subjectivity and silence that I was seeking. After a number of false starts, I found that with a discourse analysis (see Hollway, 1989; Lovering, 1994; Parker, 1992; Weedon, 1987), I could go beyond traditional attitude and categorisation research. Discourse analysis theorises subjectivity as constituted in a specific social world through power relations and argues that we do not and cannot have direct access to the 'truth' of either lived experience or the material body *outside* of discourse or history.

Discourse Analysis: Girls' Talk about Menstruation

Growing up Female

The problematics of the menarcheal body was first signalled in the interviews by the girls' ambivalence about 'growing up'[2] and their silence about the inevitable physical developments of puberty. On the one hand, the girls talked about growing up as important and 'exciting' with the promise of new possibilities and pleasures. (As we will see later in this chapter, the importance of growing up was a recurring theme in these interviews.) On the other hand, they spoke with regret of their loss of childhood freedoms as they matured. At the same time, there was another aspect to growing up that the girls did not mention and when questioned found difficult to discuss: their developing bodies and starting to menstruate.

It was physical and sexual development which turned out to be more problematic and difficult than the girls had anticipated or acknowledged. When asked if they were looking forward to the physical changes associated with growing up, at best they expressed a resigned acceptance of the inevitable or at worse they said they were not looking forward to their bodies changing. A few just avoided the issue saying: 'Don't know, [I] haven't thought about it'. As if somehow if they didn't think about *it*, *it* wouldn't happen to them, but they were never explicit about this rejected '*it*'.

The 'it' girls really didn't anticipate or care to acknowledge was menstruation. It was rarely or spontaneously spoken about and never desired. As a girl said: 'Mmm ... sometimes I wish I was a boy like because they don't have all *these things* that we do'. It was not that other aspects of physical development were not problematic: like developing breasts, body size or appearance in general. It was just that the most consistently difficult aspect of the developing female body for these girls was menstruation and reproduction.

There were several aspects to this difficulty. One aspect was the patriarchal discourse about the female body as extraordinary and fascinating, the object of the male gaze and commentary. It was girls' experience that whenever menstruation or reproduction was mentioned in school when boys were present, the boys would tease the girls. As the girls said, the boys were always 'mucking around about it', sniggering and laughing and 'running round going have you had your period yet'.

The Regulation of Sexuality

Another aspect is the way the regulation of sexuality and the disciplining of the body (Foucault, 1978) results in young adolescents having restricted access to knowledge about and an understanding of their own or the opposite sex's body. In everyday life the actual bodies and sexuality of adults and children are covered up or confined to the privacy of bathrooms and bedrooms. Most young adolescents do not see naked female or male bodies, babies born or breastfeeding, men's semen or women's menstruation, or a couple's lovemaking. Neither are any of these topics openly and explicitly discussed in front of the children. The covert and overt message to young adolescents is that real bodies are abject in some way and neither to be uncovered nor spoken about.

This regulation of and silence about the reproductive and sexual body in general and female menstruation in particular leaves young people poorly informed about the realities of the menstrual cycle. It was this aspect of menstrual discourse and practice to which a few girls objected saying 'Oh ... [the worst thing is] just having to keep yourself to yourself' and 'it's a subject that's always like been 'they shouldn't talk about it in front of people [but] I think you should get it out in the open a bit more'. There were girls who wanted this process to be more open, in part so that they would not have to be so guarded and careful themselves, in part so that the males in their lives would tease them less and be more understanding and in part so that they would just know more.

This is a major contradiction in the inscription of young adolescents into the discourses and practices of menstruation. On the one hand, the menstrual period is constituted as a private, female hygiene matter concealed in public and confined to bathrooms and surgeries. On the other hand adolescent girls need to know enough about menstruation to learn to hide it and adolescent boys to keep quiet about it. The silence has to be broken, the reproductive body has to be uncovered and menstruation has to be revealed. But when and how and by whom and to what effect? It appears that boys hear enough to tease, taunt and engage in a 'sexist discourse' (Walkerdine, 1990: 5) which asserts their difference from and power in relation to girls, but they don't learn to respect or understand what girls 'have to put up with'.

Girls Learning about Menstruation

And girls, what do they learn about menstruation? First, it is important to realise that for the majority of adolescents in Western culture, the dominant discourse and practice of menstrual silence and concealment in everyday life has been so successful that the majority of girls know little about it before they are told. In this study, most of the girls never appeared to have seen any concrete evidence of women's bleeding, not even their mothers' with whom they lived day in and day out, nor to have heard any explicit talk about it. It was not until they were nine or 10 years old that they were taken aside to be told about it and most 'didn't believe it'. Indeed, it must be hard to believe. One day more or less out of the blue, having seen and heard next to nothing, girls get told that women bleed from down there.

Having had little idea that such a thing happened, the girls had difficulty believing that women in general have periods or that they in particular would. They were 'shocked' saying 'it doesn't sound very nice', 'I thought, Oh, No!', or 'I thought yuk!' Not one girl in this study expressed any positive reactions to learning about menstruation. And of course, a number worried about this sudden and unexpected bleeding: 'Um ... I wasn't very happy, I don't know ... thought it would hurt or something'.

Although most girls remembered both their reactions and feelings on learning about menstruation, there were girls who said that they couldn't remember anything or that they had felt 'nothing'. One girl went so far as to refuse to know about menstruation at all. How is such a refusal to be explained? One approach could be to place the girl's response into a category such as 'don't know' or 'silence'; a woman-centred approach. Another approach could be to utilise the concept of denial or repression; a psychoanalytic approach as suggested by Helene Deutsch (1946). A third approach could be to consider this an empirical problem; a failure in the interview method. The difficulty with any of these explanations is their failure to address the impact of the social on the subjective. Girls' responses to menstruation are not *just* a reflection of their unmediated real experiences or their psychic processes. Girls' responses to menstruation are a reflection of present patriarchal menstrual discourses and practices which I suggest are about the contradictory processes of concealing menstruation, constructing it as a pathological female process and refusing it a place in everyday life while at the same time wanting women to desire femininity, maternity and normality.

This places girls in the contradictory position of needing to know about something that people don't talk about, women don't appear to have and society doesn't value. Is it surprising that girls did not think much about menstruation or refused to learn about it? Although it is possible for menstruation to be a traumatic event, for mothers to inadequately prepare daughters or for schools to fail to teach about menstruation properly, I am suggesting that these girls' responses to menstruation were the result of more than any of these factors. The girls did not remember about or have many positive responses to menstruation because they were not encouraged to be positive about it.

This was reflected in the contradictions of the girls' education and knowledge about menstruation. Although the schools taught about menstruation, most girls viewed their education at best as 'a bit' in 'science' about which the boys teased them. There are several points I want to make about this teaching. First, whatever it was that was being taught about menstruation in school, it was not relevant enough for the girls to consider that they had been taught very much at all. The scientific discourse of menstruation, even when taught by a woman, was for these girls about school, reproduction and the female as Other; not about their potential or actual menstrual cycle.

Second, when menstruation was taught to girls either by men or with boys, there was always the potential for, or the reality of, the teasing, sniggering and embarrassment about the female body and menstruation because menstruation was constituted as an extraordinary female process linked to sexuality and fertility. At this age and in a public space, it was not a process that girls even wanted to hear about, let alone learn about or ask questions about. As one girl responded when I asked 'Well, if you don't know very much, why do you think you know enough about it then?' Girl: 'I don't really want to know anything about it'.

This leads me to the next contradiction. The majority of girls considered that their mothers were the most appropriate source of information and should 'tell' them about 'it'. However, the dominant discursive practices of menstruation have not encouraged women even if they are mothers of daughters to discuss or feel positive about the menstrual cycle, let alone prepared them to teach about it to a young adolescent. If a woman considers her own menstruation as normal and pain free, what is there for her to say? If a woman considers her own menstruation as difficult and painful, what should she say? If a woman understands menstruation as linked to adult sexuality, when should she talk to her child? And all of this is to be done within an idealised mother/daughter relationship: a mother/daughter pair that even girls realised can have difficulties and limitations. As the girls pointed out: Girls '*should* be able to confide' in their mothers, but of course not all girls can. And mothers *should* be able to comfortably tell their daughters, but of course not all mothers can.

Premenarcheal Girls' Descriptions of Menstruation

So how do premenarcheal girls describe and explain menstruation? The most frequent explanations of menstruation drew on two related but different discourses. One is the discourse of blood loss. Here girls described menstruation as: 'Just blood that's left over in your body ...', or you 'lose all the blood'. Of course for some girls this idea of losing blood was frightening because in our culture blood is associated with injury and illness. However, blood is also associated with life and vitality so there are alternative and older theories about the menstrual flow available to make sense of menstruation. These ideas were hinted at in these girls' explanations of menstruation as 'Just blood that's left over ...', or 'too much blood [that's] not being used up ...'. In these theories the menstrual flow is a sign of health: it is excess to a woman's requirements and a cleansing process. Here menstruation is constituted as a normal, healthy process.

The other, more dominant discourse was the more scientific discourse of the reproductive cycle in which menstruation is caused by failed reproduction, it is a sort of waste-product and linked to fertility. However, this link with fertility and hence sexuality was not a possibility girls welcomed at this age. In their explanations of why women menstruate, girls talked about 'when *they* get older men fertilise it' and '*they* have children': they tended to distance themselves from this unsought fertility. Certainly as other researchers have found menstruation was a sign of *potential* fertility (Apter, 1990; McGrory, 1990), but the majority of girls did not talk about it as the symbol of fertility or the sign of womanhood. Rather they talked about the onset of menstruation as a sign of 'growing up'. This was the only aspect of menstruation that premenarcheal girls thought they would like.

So what did girls actually expect the onset of menstruation to be like? From the number of 'Don't know' responses, I suggest that any explanations of menarche prepared these girls for their first period in only the most basic of ways. It appeared that they had not been given explicit or what could be called phenomenological descriptions of menstruation (Martin, 1989). They knew what they had to do, use sanitary towels, but they didn't know what to expect.

The lack of explicit information in conjunction with the knowledge that menstruation entailed bleeding both scared and worried girls because as they said: '[It's] a bit scary because I've never had it before and you don't know what it's like'. A number thought there might be pain or discomfort: 'Mmm, it will be a bit uncomfortable at first' or 'A bit sore and painful'. In addition there was the more worrying possibility of embarrassment. Their first period might happen in a public place catching the girl unaware and unprepared. As one girl explained:

> I think it will be dead embarrassing if it happens somewhere that, it would be okay if it happened in the house but if it happened somewhere like in the pool it would be dead embarrassing.

This was one of the unspoken fears of starting menstruation, suddenly having blood everywhere, and not just that but in a public place. In the absence of detailed information about how much blood there might be or how much warning they might have, there was a fantasy about being flooded with blood for which one was unprepared; unprotected by the bulky but reassuring sanitary towels. It may have been this unspoken eventuality that led girls to expect the first and therefore, unprotected menstrual flow to be:

> Horrible! Well, I don't know actually. Well, it can't be that bad really, but it won't be very nice, not the most enjoyable thing.

It also led to a few girls staying in a state of constant readiness. As one girl said:

> Every time I go to the toilet I look, thinking it has happened. I don't know what it is going to be like, I don't know.

It was no wonder that these girls were finding their developing bodies a source of ambivalent feelings and difficulty. Neither was it any surprise to learn that none of the girls was positively looking forward to the onset of menstruation. At best a few girls wanted their periods to start because it was a sign of growing up or to relieve the anxiety of waiting. At worst, many did not appear to want them at all: 'Well, I wouldn't have them at all. Really, I could do away with them and just say now produce an egg'.

There were few advantages for these young adolescent girls in beginning to menstruate. They did not expect menarche to have any effects on their relationships with other people nor to bring them any particular privileges, benefits or celebrations. They did not even expect to tell many people when they began menstruating, only their mothers. And there were a few girls who said they would tell no one, not even their mothers.

A major concern for these girls was that their menstruation not show and not be 'broadcast'. Whatever else happened, they hoped they would begin menstruating at home, preferably and literally 'on the toilet'. Girls worried 'about when I'm going to have it and where' preferring it 'didn't happen at school' or in any other public place because 'then it will be embarrassing'. The practice of women being silent about and concealing their menstruation in everyday life was so dominant for these girls that it had taken on the nature of an imperative.

Postmenarcheal Girls' Accounts of Menstruation

And finally, considering premenarcheal girls' explanations of menstruation and expectations of menarche, how did postmenarcheal girls understand and explain the onset of menstruation? There were two major themes in the postmenarcheal girls' accounts of starting menstruation: its management and its insignificance.

In terms of management, whatever else the girls included or left out of their accounts of their first menstrual period, they nearly always talked about whether they had been at home or school. If they were at home, they could change their clothes and obtain sanitary protection relatively easily. In addition their mothers were more likely to be available for practical assistance and reassurance. However, if they were at school without their own sanitary towels or tampons, they had two problems. One was the necessity of actually obtaining sanitary protection and the other was being embarrassed by some revealing stain. The enormous relief of girls who started to menstruate at home was in marked contrast to the distress of unprepared girls who started at school. This made the frequent comments about not noticing or not remembering the first menstruation all the more surprising.

This leads me on to the other major theme: the lack of significance and impact of menarche. When asked about their first menstruation, a number of girls simply replied: 'I don't remember, I started at school, I remember that', 'Didn't really notice', or even 'I don't know'. These were puzzling responses to an event which some girls in this study described vividly.

In looking at their accounts, there were two aspects to this failure of memory and attention. One aspect was the lack of any dramatic element in the actual physical event of beginning the menstrual flow. There was no pain, no sudden gush of blood and no strong physical sensation so that in essence they 'didn't notice' the beginning of menstruation until they found a stain in their pants.

The other aspect was the lack of reaction by others and subsequently the lack of effect on themselves. When they told someone, primarily their mothers, about beginning menstruation, the common reactions were described as practical and low-key. Girls remembered being shown what to do and told not to worry. Fathers were not involved in any direct way: most girls did not tell their fathers or other family members with the majority not even knowing if their fathers knew they had begun menstruation.

The majority of these postmenarcheal girls said that people treated them no differently than before their menarche, there had been few effects on their activities and they felt no different except for feeling slightly more grown up. The people around them seemed neither interested in nor concerned about their menarcheal experience. In essence these girls were not remembering or noticing their menarche because others indicated there was nothing of significance or importance to remember or notice. As two girls said:

> ... Not very ... like everybody talks about it being really like different from then on but it wasn't really.
>
> Oh, sometimes, I just think that everybody will know. Just, yeah, just walk and everything and you think everybody's going to know this or something like that, but nobody said, nobody noticed at all.

In contrast to the girls who forgot, didn't know or didn't notice, I want to discuss two girls who gave quite a different account of the possible impact and significance of menarche. The first shows the possible effect on girls of having the opportunity and encouragement to talk about their expectations and experiences of menstruation. This girl's menstrual periods had started soon after I had first interviewed her and then in the second interview when asked about menarche the first thing she said was: 'It was just after you talked to me. *It was easier*'.

My point is not that talking about menstruation necessarily removes the surprise or even the fear, but that the silence about and concealment of menstruation fails not only to prepare girls for the actuality of starting their menstruation but to give their experience any significance. If no one appeared interested in what girls might remember or say, then why should they remember, notice or say anything? When alternative practices and discourses gave impact and significance to menarche, then girls' accounts were quite different. This was particularly striking in the second girl's account. Here the practice of publicly celebrating a girl's menarche had made a strong impression.

When this girl talked about her first period, she launched into a relatively long and full account of a girl's celebration which she had witnessed. Here are her comments:

> [My first period] It was like exciting first of all, you know, but I asked my mum [not to do the same]. Because when my friend started her period in France last year, they were all like clinking their glasses and they were really congratulating [my friend]. But I said I didn't [want that] because that would embarrass me ... because we were all just sitting around the table and then um Jolly like, she goes, argh, like this and she told her mum. Then Gloria told my mum and they they were all going, ooooohh, like this and then they were all getting wine out and everything.
> (Interviewer: And your own menarche?)
> [I told] My mum ... just my mum.

This public celebration of menarche made an impact on this girl's menarcheal experience even though she didn't have a celebration herself: she was the only girl to describe her first period as 'exciting first of all'. The excitement of her friend's celebration imbued her own menarche with some measure of excitement and significance. It certainly gave her something to say and something to remember about it as an event, even if her own more private menarcheal experience was subdued.

But aside from these exceptions, most girls simply had little to say about their experiences of menstruation. This has been noted in other studies with adolescent girls (Apter, 1994). However, there was one aspect of menstruation about which (some) girls had more to say: menstrual distress and pain. This was the one discursive practice easily available to the girls and to me. Whereas women and girls are expected to be silent about and conceal menstruation and whereas there is no discourse of menstrual pleasures and desires, there is a medical discourse of menstrual pathology and a discourse of menstrual disgust in which the pains, aches, moods and miseries of menstruation can be discussed.

In making this point about the discourse of menstrual pathology, I am not trying to suggest that girls suffer from menstrual distress because they are unprepared for menstruation, have negative attitudes to menstruation or expect to have menstrual difficulties. In fact these interviews suggest there is no such straightforward relationship between expectations about or preparation for menstruation and menstrual attitudes or symptoms. There were girls who expected pain, but did not have any. And girls who didn't expect pain, but did have it. In the event, in this study half the postmenarcheal girls did report some menstrual symptoms saying they got 'grumpy before' or 'sometimes I have tummy aches' but girls tended to describe these symptoms as occasional, brief or transitory and they did not seek treatment. They viewed them as part of *normal* menstruation and consequently they had little to say about them. Whereas the few girls who were experiencing more severe problems talked more expansively about their menstrual problems than other girls did about any other aspect of menstruation.

Conclusion: The Problematics of the Menarcheal Body

What I am arguing on the basis of this research is that the dominant discourses and practices of menstruation construct menstrual experience so that there is little that is positive about menstruating for adolescent girls. A few postmenarcheal girls liked the feeling of growing up; however, most girls, especially when they were postmenarcheal, could find nothing they liked about menstruation. Depending on their menstrual symptoms and desires, they found it more or less restricting, inconvenient, uncomfortable and distasteful. In the absence of a discourse of menstrual pleasures or a practice of menstrual celebration, the onset of menstruation had, as the girls expected, brought them few advantages and fewer delights. There was not even much to talk about except aches, pains and mess. The only advantage of menarche was that it signified growing up and this was important. For slightly more than half the girls, it was important enough that they would not stop menstruating given the chance. At least they felt they were developing normally, as they said:

> Well, you just, you know, if you don't have them, people would think there was something wrong with you or something.
> It would seem strange [to stop menstruation] like you weren't growing up or something.

However, for a substantial number of girls the importance of growing up and normality was not enough, there were too many hassles and too few pleasures so that they wished to menstruate only 'when I wanted to have a baby'.

I suggest that what this research shows is not so much the particular significance of social science discourses of menarche nor of the actual onset of menstruation on girls' subjectivity, but the dominance of discursive practices which refuse the woman's body and functions a place in the everyday social world. Although menstruation and menarche are and have been *interminably* discussed in the academy and in the surgery, they have been constituted as the objects of expert scientific, psychomedical and pedagogical practice which signifies the female body as 'other' and menstruation as on the borders of pathology. They are not part of men's and women's everyday conversation, visible practices or mutual arrangements. It is within these dominant discursive practices that girls and women have to function and from which they have to constitute menstrual knowledge and meanings.

Therefore, we cannot afford to view menarche and its attendant difficulties as arising out of the physical realities of girls' developing bodies, their lack of preparation or poor social education. Such constructs both belie the immense complexity of the discourses and practices of the female body and menstruation and explain it away by making it mothers' or educators' problem. By ignoring the cultural and historical construction of the body, a particular historically produced female body discourse and practice can remain intact.

The girls in this study were *failing* to learn about menstruation, they were *failing* to be prepared for menarche and they were *failing* to have positive attitudes to menstruation. But this was not because their mothers or even their teachers were ignorant or inadequate, or just because boys teased, or even because bleeding genital orifices were inevitably disgusting or frightening. Neither telling girls about the mechanics and management of menstruation nor teaching them positive attitudes is going to solve the difficulties of becoming a menstruating adolescent girl.

There is no easy womanhood, no simple adjustment to sexual maturation thanks to the sensitive mother or enlightened educator preparing the way through the crisis inside. Any failure of girls to adjust to the onset of menstruation and any crisis of girls on reaching menarche is an aspect of the historical construction of a female sexuality and silence with menstruation as the event *par excellence* which symbolises women's difference and fate. This is the powerful regulation of the feminine.

This does not mean there is nothing to be done to improve girls' lives. There are some simple things that the girls in my research suggested themselves, as have girls in other studies (see Holly, 1989). And it is not menstrual celebrations: the danger here is the re-production of the female body and nature as essentially and incommensurately different from the male. Girls wanted a combination of more equality between boys and girls with some respect for and recognition of sex differences. These girls suggested that girls and boys be given their education about puberty, sexuality and menstruation in single sex groups at some stages and ages, in mixed groups at other times; that any discussions and educational materials be equally about girls and boys, their bodies and their role in reproduction; that women teachers be available to teach these areas to girls and men teachers to boys; and that both primary and secondary schools set up mechanisms that accept menstruation at a public level, including decent toilet facilities, easy access to sanitary products and recognition of girls' menstrual requirements as well as the reality of menstrual discomfort so that girls do not have to do school sports in shorts with bulky pads or with menstrual pain.

If, in addition to these simple things, schools were to take up what Jacqui Holson considers their duty 'to challenge and condemn boys' and men's' sexual violence and, thereby, to help empower young women to challenge and change their world' (1989: 141) and if developmental psychology as well as other social sciences were to become involved in what Ben Bradley (1993: 412) considers 'an increasingly profound and intimate way with the human struggle for liberty and equality', then there might be a better possibility for social justice for adolescent girls.

Certainly when my research into the historical construction of the female body and my discussions with young adolescents are collected together they tell painful truths about how boys and men taunt, embarrass and denigrate pubescent girls; how girls are negative, anxious and ill-informed about menarche and menstruation; how mothers, teachers and health educators fail to impart positive attitudes to, dispel fears about, or ensure good experiences of menstruation; how social practices and public institutions refuse the female body and functions value or place; and how scientific, medical and pedagogic discursive practices construct the female body and sexuality as the extraordinary 'other' bordering on pathology. It may be that these truths will be difficult to accept and will be met defensively by parents, professionals, public officials and scientists. The dominant discourses of the female body and the development of adolescent girls have been remarkably resistant to any fundamental change. The language may evolve and new facts may be discovered but the story stays the same; women are essentially and incommensurately different from men by virtue of their embarrassing, bleeding bodies.

It has not always been like this. At an earlier point in Western culture, menstruation was once the 'flowers' (Sharp, 1671: 288) and menstruating women were desirable and desirous (Lovering, 1994). Women have resisted the transformation of the female body into a somatised and pathological object of the male gaze and practice. We have created alternative visions of the healthy and capable female body from the Victorian period (see Jacobi, 1878) to the present (see Martin, 1989; Nicolson, 1992; Sayers, 1982; Walker, 1992). If it has been different before then it could be different again, but to be different the practices of the regulation of the feminine must be transformed. We need different stories to help in the transformation of the discursive constitution of the feminine body.

Girls in this study give us some hope for this transformation. They refused to learn the correct scientific theories, they complained about menstrual silence and they suggested different educational practices. We need to continue this resistance to the same old story. We need to tell a different tale. For feminist psychology and science I suggest incorporating into our research an understanding of the history of social forces that produce ideas of the female and male body, femininity and masculinity and the psychology of women and menstruation. For women I suggest we need to talk about menstruation with friends, colleagues, lovers, family, children and adolescents. This is a necessary (and I know a difficult) step even if it is not a sufficient step in any transformation of sexist patriarchal discourses and practices of menstruation, menarche and psychology. The girls in my study encourage us by saying:

> It's a subject that's always like been, they shouldn't talk about it in front of people, I think you should get it out in the open a bit more.
> I think it's important to find out about ... talk about growing up. Um ... I thought it would be embarrassing but it wasn't – I could talk and everything.

Notes

1. The interviews were done twice in two years with 46 girls from South West Britain. They were predominately white European and drawn from across the academic ability range and social classes. The interviews were part of a larger study on young British adolescents' attitudes to and experiences of growing up, in particular the physical aspects of development. The study began when the young people were 11 to 12 years old and had three parts: a series of discussions with three groups of girls and three groups of boys done once, psychometric tests and questionnaires with 80 girls and 80 boys done at three points, and the interviews with the 46 girls done twice. See Lovering (1994) for further details.
2. All text in quotation marks not attributed to an author in this section are direct quotations from the girls' interview transcripts and all italics are mine for emphasis.

Acknowledgements

My special thanks to Avril Butler, Michele Cohen and Anne Walker for their helpful comments on a draft of this chapter.

References

Amann-Gainotti, M. (1986), 'Sexual Socialization During Early Adolescence: The Menarche', in *Adolescence*, vol. 21, pp. 703–710.

Apter, T. (1990), *Altered Loves: Mothers and Daughters During Adolescence*, U.K.: Harvester Wheatsheaf.

Belenky, M.F., Clinchy, B.M., Goldberger, N.R. and Tarule, J.M. (1986), *Women's Ways of Knowing. The Development of Self, Voice and Mind*, New York: Basic Books.

Bradley, B.S. (1993), 'Introduction: The Future of Developmental Theory', in *Theory and Psychology*, vol. 3, no. 4, pp. 403–414.

Deutsch, H. (1946), *The Psychology of Women. A Psychoanalytic Interpretation, Vol. 1 Girlhood*, London: Research Books Ltd.

Foucault, M. (1978), *The History of Sexuality: An Introduction*, U.K: Penguin, translated from French by R. Hurley and reprinted by Peregrine Books, 1984.

Gilligan, C. (1982), *In a Different Voice: Psychological Theory and Women's Development*, Cambridge, MA.: Harvard University Press.

Grief, E.B. and Ulman, K.J. (1982), 'The Psychological Impact of Menarche on Early Adolescent Females: A Review', in *Child Development*, vol. 53, pp. 1413–1430.

Hall, G.S. (1904), *Adolescence, Its Psychology and Its Relations to Physiology, Anthropology, Sociology, Sex, Crime, Religions and Education*, 2 vols, New York: D. Appleton and Co.

Henriques, J., Hollway, W., Urwin, C., Venn, C. and Walkerdine, V. (1984), *Changing the Subject: Psychology, Social Regulation and Subjectivity*, London and New York: Methuen.

Hollway, W. (1989), *Subjectivity and Method in Psychology: Gender, Meaning and Science*, London: Sage.

Holly, L. (ed.) (1989), *Girls and Sexuality*, U.K.: Open University Press.

Holson, J. (1989), 'The Sexual Harassment of Young Women', in Holly, L. (ed.), *Girls and Sexuality,* U.K.: Open University Press, pp. 130–142.

Jacobi, M.P. (1878), *The Question of Rest for Women During Menstruation*, New York: G.P. Putnam.

Lee, J. (1994), 'Menarche and the (Hetero)Sexualization of the Female Body', in *Gender and Society*, vol. 8, no. 3, pp. 343–362.

Lei, C., Knight, C., Llewellyn-Jones, D. and Abraham, S. (1987), 'Menstruation and the Menstrual Cycle: Knowledge and Attitudes of Mothers and Daughters', in *Australian Journal of Sex, Marriage and Family*, vol. 8, pp. 33–42.

Logan, D.D. (1980), 'The Menarche Experience in Twenty-Three Foreign Countries', in *Adolescence*, vol. 15, no. 58, p. 287.

Lovering, K.M. (1994), 'Discourses of Menstruation: Girls, Menarche and Psychology', unpublished PhD thesis, University of London.

Lovering, K.M. (1995), 'The Bleeding Body: Adolescents Talk About Menstruation', in Wilkinson, S. And Kitzinger, C. (eds), *Feminism and Discourse*, London: Sage.

Martin, E. (1989), *The Woman in the Body*, Milton Keynes: Open University Press.

McGrory, A. (1990), 'Menarche: Responses of Early Adolescent Females', in *Adolescence*, vol. 25, no. 98, pp. 265–270.

Nicolson, P. (1992), 'Menstrual-Cycle Research and the Construction of Female Psychology', in Richardson, J.T.E. (ed.), *Cognition and the Menstrual Cycle*, New York: Springer Verlag.

Parker, I. (1992), *Discourse Dynamics: Critical Analysis for Social and Individual Psychology*, London: Routledge.

Potter, J. and Wetherell, M. (1987), *Discourse and Social Psychology: Beyond Attitudes and Behaviour*, London: Sage.

Rierdan, J., Koff, E. and Stubbs, M.L. (1989), 'Timing of Menarche, Preparation and Initial Menstrual Experience: Replication and Further Analyses in a Prospective Study', in *Journal of Youth and Adolescence*, vol. 18, no. 5, pp. 413–426.

Rose, N. (1985), *The Psychological Complex. Psychology, Politics and Society in England 1869–1939*, London: Routledge and Kegan Paul.

Ruble, D.N. and Brooks-Gunn, J. (1982), 'The Experience of Menarche', in *Child Development*, vol. 53, pp. 1557–1566.

Sayers, J. (1982), *Biological Politics*, London: Tavistock.

Scott, C., Authur, D., Owen, R. and Panizo, M. (1989), 'Black Adolescents' Emotional Response to Menarche', in *Journal of the National Medical Association*, vol. 81, pp. 285–290.

Sharp, J. (1671), *The Midwives Book*, London.

Skandhan, K.P., Pandya, A.K., Skandhan, S. and Mehta, Y.B. (1988), 'Menarche: Prior Knowledge and Experience', in *Adolescence*, vol. 23, pp. 149–154.

Squire, C. (1989), *Significant Differences – Feminism in Psychology*, London: Routledge.

Walker, A. (1992), 'Premenstrual Symptoms and Ovarian Hormones: A Review', in *Journal of Reproductive and Infant Psychology*, vol. 10, no. 2, pp. 67–82.

Walkerdine, V. (1990), *Schoolgirl Fictions*, London: Verso.

Weedon, C. (1987), *Feminist Practice and Poststructuralist Theory*, Oxford: Blackwell.

Weiden, J.P. (1984), 'Menstrual Attitudes, Self and Body Image and Overall Psychological Adjustment in Pre- and Post-menarcheal Girls', PhD dissertation, University Microfilms, Adelphi University.

Reading the Silent Body

Women, Doctors and Pelvic Pain

Victoria Grace

> The body and the mind do so reciprocally influence each other that we can scarce keep the one in tune if the other be out of it.
> (Mary Astell, 1704)

It is not uncommon for women consulting the health services of Western medicine for chronic pelvic pain (CPP) to experience not only a failure to receive a diagnosis and treatment for the symptoms, but also a process which systematically undermines their self-confidence and their ability to control and manage their pain and their lives. My intention is to present a theoretical analysis of the medical construct of the physiological and the psychological in staging this scene.

I have recently completed a research project concerned to understand women's experiences of using the health services for chronic pelvic pain (Grace, 1995a and 1995b). This issue first came to my attention as a story many women with chronic pelvic pain tell: one where they have terrible pain, go to the doctor (general practitioner or gynaecologist) probably over and over again, repeatedly feeling somewhat dismissed when their pain is not, it seems, taken seriously, or when they are faced with the implied or explicit assessment that it is 'all in your head', that they are 'neurotic'. They do not have a diagnosis, or more importantly, they do not have an understanding of their pain. Frequently they have no treatment, or inappropriate treatment and often are diagnosed many years later with endometriosis, or with some other condition that might be able to be successfully treated medically. But there are many instances where the woman does not receive a diagnosis, even after years of consulting doctors, where, no matter how hard they look, they cannot 'find anything'.

In the early stages of the research I did a qualitative discourse analysis of women's narratives on their experience of consulting the medical profession for chronic pelvic pain and what emerged from that analysis was a clear picture of the medical reduction of the complexity of meanings of pelvic pain to the physiological, to that which can be *seen* with the eye of clinical anatomy (Grace, 1995a). The woman wants to tell the doctor about her pain in an endeavour to have the reality of the pain acknowledged in that encounter. Rather than encountering a doctor who enters into a dialogue about the meaning of her symptoms she confronts a doctor who is somewhat dismissive of her words and who extracts her symptoms by a variety of manual and technological means, constructing symptoms in medical terms and reducing the diagnostic procedure to one whereby the entire focus is on whether or not 'something is found'.

'Nothing was found' becomes the worst possible outcome of a diagnostic investigation conducted on the basis of this medical premise. A premise whereby the reality of the pain is contingent on whether or not something was found; a kind of crude, fixed materiality of what is ultimately a visual evidence required as proof of pain and required for knowledge of what 'causes' pain. And without that evidence, without the presence of an organic pathology, if the cause is not manifested in the physiology of the organism, it must therefore be 'psychological'. Implications that the woman is 'neurotic' are as much a part of her cultural imagery as they are of the doctor's: that the pain is not really real.

Chronic Pelvic Pain Without Organic Pathology

An analysis of the English language literature on CPP revealed a fascinating construct: 'chronic pelvic pain without organic pathology'.[1] How is the question of the physiological and the psychological discussed in this medical, psychiatric and psychological literature? The English language literature appears to be dominated by contributions from North America, and in North American journals. This is augmented by contributions from Britain, France, Denmark, the Netherlands and to a lesser extent Italy, Germany and Australia. But even so, the established dominant conceptual frameworks of North American medicine and psychiatry appear to have set the terms of the debate. It is interesting that an important aspect of this framing of the terms of the debate(s) is the very term 'chronic pelvic pain without organic pathology', a term defined in the negative.[2] If there is no organic pathology present, or 'found', then the attention turns to the 'psychological' arena.

The medical and psychiatric literature on CPP emerging in the late 1940s and moving through into the 1960s and 1970s was dominated by what I would call a pseudo-Freudian understanding that constructed 'women's psychology' in terms of assumptions about 'women's nature'.[3] The literature is replete with examples of this ideology:

> The patient with pelvic pain was typical in her angry, attacking and undermining attitude towards her husband and was quite unable to view her marital problems except with reference to his shortcomings ... the patients with pelvic pain had unusual difficulties in the establishment of a mature feminine identity [which] is related to an unresolved ambivalent attachment to the mother and to underlying confusion and anxiety about female anatomy ... they were dissatisfied with being housewives and cherished fantasies of going to work.
>
> (Gidro-Frank, Gordon and Taylor, 1960: 1191)

> Pelvic pain patients ... [showed evidence of] serious emotional deprivation and lack of love ... an example of the futile and unsuccessful use of genital experiences to resolve pre-genital difficulties.
>
> (Castelnuovo-Tedesco and Krout, 1970: 124)

> Often these women develop their pain when life is going well for them. They need to suffer because of anger and resulting guilt. If the environment is difficult enough, they may not need physical symptoms.
>
> (Friederich, 1976: 402)

> These (pelvic pain) patients have tried to overcome emotional deprivations in childhood and need for closeness and to solve their problems about dependency relatedness and trust by getting involved in incestuous relationships ...
>
> (Gross *et al.*, 1980–1981: 93)

This kind of analysis of the 'pelvic pain patient', especially without organic pathology, dominated the literature for decades. It was somewhat in tension with a much smaller group of authors who rejected the entire notion of 'without organic pathology' in relation to CPP. This latter approach remains firmly within the duality of physiology/psychology, but rejects or denies the possibility of psychological involvement in the condition. The pain *is* the result of tissue damage that exists but has not yet been found. Therefore, it is a matter of looking more vigorously and indeed, increasingly hidden and mysterious physiological causes have been manifested, or literally brought to light through the use of the laparoscope.

The other point of tension with the dominant psychoanalytic interpretations within the literature was with those who argue that psychological pathology is rather the effect of the pain and not the cause. This debate was quite active for a brief period, but did not really gather momentum, possibly because the 'cause' remained elusive regardless and the criticism did not appear to make much difference to the overall discussion of the issues.

In the 1970s, 1980s and into the 1990s, the North American psychoanalytic discourse gradually becomes replaced by, or influenced by a different kind of talk about 'women's psychology', one employing cognitive theories of psychology and social learning theory, or behaviourist psychology. Latterly we see the following kinds of statements: 'women suffering from chronic pelvic pain feel very insecure and may have conflicts in gender identity' (Bekker, Peters and Bayer-Rots, 1990: 19). Pelvic pain patients are contrasted to controls, who are described as 'normal, average women with respect to gender identity development' (Bekker, Peters and Bayer-Rots, 1990: 25). We see reference to 'psychiatric morbidity and illness behaviour ... depressed mood' (Hodgkiss and Watkins, 1994: 3); 'sexual abuse', 'sexual problems', 'neurotic behaviour', 'psychological disorders' (numerous papers).

In addition, authors invariably use numerous different scales and measures to develop profiles of the psychology of women with chronic pelvic pain without organic pathology. It could be argued that there are in fact many similarities between the earlier Freudian-influenced and the contemporary psychological discourses, but I do not want to pursue that at the moment. What I want to draw attention to is the way that attempts by medical and psychiatric (and more recently psychological) researchers to understand the 'psychology of women with chronic pelvic pain without organic pathology' seem to have (a) done very little at all to help women (in fact from the research I have done in New Zealand, I would say it has done absolutely nothing to help doctors engage more adequately with this presentation of symptoms) and (b) contributed to a construction of the absence of organic pathology as somehow indicative of that which is unrepresentable in medical terms.

The binary structure of physiology and psychology within this medical frame renders that which can be identified in positive terms, empirically, as the dominant term:[4] tissue damage and named, known disease processes. That which is not able to be so identified, is the subordinate term, the absence, the lack, the *without*. And indeed, in medical terms, is not knowable. The attempt to *know* in psychiatric or psychological terms has not produced *knowledge* that is of use to the medical person, nor for that matter of use to women either as patients or independently from the medical consultation. The project of the literature is to *assert* a positive knowledge (in the tradition of Western empiricist science) of the psychology of women with this 'condition', but its results, it seems, cannot be activated in any meaningful way; in the medical context such a 'condition' remains in the order of the unknown and unknowable.

The woman with CPP is thus configured as an unstable and intensely problematic site. She is both an empirical body that is knowable and representable through the gaze of clinical anatomy and a speaking subject whose agency haunts the clinician's inability to see/know. I will analyse the two interrelated aspects of this construct: the primacy of the visual and the silencing of the female subject.

Visualisation and Medical Ontology

Feminist analysts have highlighted the way the intensely gendered dominance of the visual sense acts as a marker for reality in Western culture(s). Luce Irigaray, in particular, has challenged the psychoanalytic assumption of sight being the primary anchoring point for the oedipal drama. More specifically, Rosi Braidotti (1994) has analysed the significance of the primacy of the visual, or 'scopic', within medical discourses in the context of a discussion on embodiment and sexual difference in contemporary feminist theory. She argues that the 'rational, visible organism of modern science marks the end of the fantastic, imaginary representations of the alchemists and consequently empties the body of all its opacity and mystery' (1994: 62).

Drawing on Foucault's depiction of the medical gaze in the contemporary era of biopower,[5] Braidotti uses psychoanalytic concepts to present an analysis of the connection between the 'scopic drive', knowledge and control, or domination. In simple terms, she describes the practice as one that involves an impulse to 'open up' for the purpose of 'looking in', to satisfy a curiosity. Within this framework the impulse to look, to lay bare, to make visible, is, she suggests, a 'fundamental and childlike form of control over the other's body' (1994: 67). Again in Braidotti's words, 'in this sense, the curiosity that pushes the child to break his or her toy to see how it's made inside can be seen as the most primitive form of sadism' (1994: 67). While the connection with a sadistic pattern is not sufficiently substantiated within Braidotti's text, it is nonetheless clear that, within the medical domain, there does appear to be a dominant 'scopic drive'. This 'drive' is necessarily implicated in the empiricist assumptions of medical science, where the other senses – auditory, kinesthetic, olfactory – interact with the visual to form the database for clinical interpretation. Rather than assuming that the dominance of the visual sense (as opposed to another such as touch) as the important pivot around which the medical practitioner's desire for control hinges, it is more fruitful, theoretically, to focus on the assumption of an empiricist ontology which takes that which can be registered by the senses as the benchmark for the real. Language then simply becomes a vehicle through which the sensually 'observed' and measured data is transmitted from one encoder to another decoder. In this process the symbolic exchange and dialogue between patient and professional is annihilated.

To the gynaecologist for whom *to know* is to identify, name, predict and control, the presence of a woman who brings to him[6] her pelvic pain, who seeks his expert help with his (and her) assumption or expectation that he has the power-knowledge to eradicate it, is possibly worse than 'perplexing'.[7] He is confronted with the realm of the unrepresentable, in the form of woman, pain, the mysterious female pelvis, absence, nothing. A kind of medical void that is anything but innocent and neutral. If the doctor defends himself against the presence of that which challenges his entire framework, if he persists in only recognising the empiricist representation of pain in the physiological organism and therefore hides behind this facade of scientific knowledge, his power will obscure the vital importance of the interactional dynamics between himself and the patient and ensure that the problem of the absence or lack is projected firmly and resolutely onto and into the body of the woman. It is almost as if her pain is doubly instituted by being recreated in the medical encounter. And conversely, the body that produces the desired knowledge (the body in which something *is found*) legitimates the power and control of the medical profession.

The progressive refinement of technological means to *see* further and further into the body may, it could be argued, intensify the medical anxiety accompanying the 'nothing found' outcome of an attempt to diagnose the source of the pain. Braidotti argues that we are in fact moving from a medical gaze that privileges visibility into a 'new culture of visualisation'. With techniques like ultrasound, that which is invisible can be technologically *visualised*, in other words can be constructed as an object of 'imaged representation'. Accordingly, the unrepresentable can finally be represented; all is revealed to the gaze, to knowledge. Within this condition of possibility Braidotti says 'the sadistic impulse of the medical gaze becomes even more of a death drive with these new visualisation techniques' (1994: 68). Braidotti gives the term 'medical pornography' to this imperative to represent living matter in an absolute transparency. Without providing greater discussion of the use of the term 'pornography', at face value this formulation could certainly be criticised. What is important for the purposes of the current discussion is the connotation of obscenity that surrounds the erasure (humiliation?) of the woman-as-subject in the process of reifying the significance of the observation of the tissues of the bodily interior. However, if one employs psychoanalytic concepts to make sense of the doctor–patient encounter (or rather non-encounter), it is important not to forget that it is always an interactive construct of demand and counter-demand: the woman actively seeks medical validation for and cure of, her pain.

Is the medical privileging of the visual (the physiological source of the truth of the mat(t)er) and the simultaneous turn to the 'psychological' in the absence of a revelation of that truth, best understood as an instance of the mind/body dualism in medicine? Although there may be good reasons for advancing this argument in the current context (the physical and the psychological certainly appear to be constructed in dichotomous terms), Braidotti's analysis, drawing on Foucault, suggests an alternative conceptualisation that possibly goes further in terms of providing an understanding of why the psychological construct is so profoundly impotent. By postulating the mind/body dualism as the structural organising principle at play, it would be reasonable to expect the mind to be the dominant term of the dichotomy. Thus, in the absence of physiological tissue damage, one would expect the technologies of psychological assessment, diagnosis and treatment to be highly sophisticated and at least to some extent successful. Instead, women are almost never referred for psychiatric, or psychological assessment and treatment and on the very few occasions when they are, such treatment appears to be based on the same empiricist principles of biomedicine and is, from the women's health point of view, ineffectual (Grace, 1995b). What appeared to be a psychological-reductionism, reverts to a biological reductionism yet again. The reference to the 'psychological' simply masks the professionals' impotence.

Braidotti situates her analysis of the (fantasy of) conflation of truth and visibility in medical and biomedical discourses within a discussion of Foucault's characterisation of modernity as precisely anti-Cartesian. In his book *The Order of Things* (1966, trans 1970) Foucault developed a detailed historical, theoretical analysis of the emergence of the (European) human subject of modernity. Modernity, he argues, is not characterised by the Cartesian dichotomy of mind (subject) and body (object), where the mind and the body are entirely separate as in the prior Classical formulation. Rather, during the eighteenth century a reconfiguration of subjectivity accompanied epistemological shifts in relation to the analysis of wealth (which became economics), of life (which became biology) and of language (which became philology). At this time, according to Foucault, the subject of modernity emerged as an *embodied subject*, in other words the body becomes itself the site of a contradiction whereby it is both the object of knowledge (clinical anatomy) and the condition of knowledge, of transcendence (the subject). The emergence of this 'strange empirico-transcendental doublet' (1970: 318) Foucault calls 'man',[8] means that two kinds of analyses came into being: those that operate within the space of the body (as the object of biological scrutiny) and those that function as a history of human knowledge – 'a sort of transcendental dialectic' (1970: 319). If the irreconcilability of these two knowledges is the signature of the modern subject as Foucault argues, then the uneasiness surrounding the continual, intensely political oscillation between them provides insight into the construction of this contradictory status of the physiological and the psychological in relation to the problematic under consideration here. It would follow that a discourse of 'it's all in your head' would be meaningless (demeaning) to women *and* an abysmal negation for the medical man.

With respect to this empirico-transcendental condition of 'man' in modernity, Foucault wrote:

> This is why modern thought has been unable to avoid ... searching for the locus of a discourse that would be neither of the order of reduction nor of the order of a promise: a discourse whose tension would keep separate the empirical and the transcendental, while being directed at both.
>
> (Foucault, 1966, trans 1970: 320)

But the empirico-transcendental body-subject Foucault refers to is, of course, a sexed body-subject (something Foucault too frequently overlooks). Perhaps, for reasons to do precisely with the gendering of the body-subject, the feminine subject is less readily a partner with the doctor in creating a discourse that would keep the empirical body and transcendental subject separate and hence is more profoundly disturbing to the medical doctor's desire/fear to see/know.

Silencing the Subject

In this context, I am interested in exploring the question of what is happening when women with CPP consult medical professionals and go through a process of feeling a loss of 'self', a loss of self-confidence ('I came out of there feeling like a nothing'), a loss of a determination and will to not be overwhelmed by the pain and disability. Is there a connection between the disturbance to the medical doctor of the unrepresentable and the silencing of woman-as-subject? How might this process be understood in a way that deconstructs the exercise of power that it represents? How is the problem of pain bound up with the possibility of the power of the medical clinician to overwhelm the agency of the person suffering?

Work by feminist sociologists Fisher and Todd (1983) and Todd (1989) working in closely-related fields, has been helpful as an initiating reference point. Todd (1989) builds on work by Mischler (1984) and also employs concepts developed by Habermas (1987). She analyses women's encounters with doctors in terms of the incompatibility of the discourse, or voice of medicine and the voice of the life-world and how the dominance of the former serves to neutralise and overwhelm the latter. But in a way Todd's analysis only scratches the surface. We need to go further in developing an understanding of the subconscious social-psychological dynamics at play that not only make this exercise of power possible, but that somehow underwrite its form.

To address the question of the role of the construct of women's (pelvic) pain as that which is, or has the potential to be, unrepresentable in medical terms and to explore the implications of the significance of representation and power, Elaine Scarry's analysis (1985) of the structure of the exercise of power through torture is in some respects illuminating.

What struck me when I started to read Scarry's work was how so many aspects of her discussion of the bodily pain of torture resonated with many of the pivotal ingredients of women's discourse on their experience of extreme pelvic pain and their gradual loss of a sense of 'self' as their pain was dismissed by the medical profession. Her statement that 'what is at stake in the body in pain is the making and unmaking of the world' is one that I could not help but recognise as relevant to this project. Her conviction that an understanding of the destruction of 'language' and 'self' need to be central to an analysis of the abuse of power through torture again, appeared to present a path worth following.

Scarry, however, fails to address the crucial question of how this combination of physical pain from the abuse of power through torture and the resulting destruction of 'language' is situated with respect to the historically and culturally specific construction of embodied subjectivity. I will position Scarry's depiction of the destruction of 'language' and 'self' through torture, within Foucault's historical analysis of the emergence of the empirico-transcendental body-subject in Western culture and will make sense of her findings within this framework.[9]

Aspects of Scarry's work, considered within these qualifications, contribute to developing an understanding of the way the context of the experience of physical pain and the construction of a sense of 'self' as a speaking subject are interconnected and how physical pain creates the conditions for social power to disengage this sense of self, in this case, women's self in the medical encounter for chronic pelvic pain. The components of Scarry's analysis involve the attributes of physical pain on the one hand and the desire for power, or will to power of a regime experiencing conditions of political instability on the other.[10] The regime that institutes the torturing of people (men and women) taken as prisoners is one that does not have legitimate authority and in the extreme case creates its power through torture: an illusion of power that is precisely the trace revealing the fact that the regime does not have power. The process of torture, in Scarry's theoretical analysis, has a logic that is not immediately obvious or apparent.

Scarry argues that the effect of torture is produced through the appropriation of the attributes of pain. Although she tends to treat the 'attributes of pain' in universal terms, this notion of appropriation leads me to question the way the 'attributes of pain' enter into the medical encounter in the case of women with undiagnosed (or difficult to diagnose) pelvic pain. The salience of what Scarry calls the 'language-destroying' attribute of pain is particularly pertinent in the context of the silencing of women's own words in the medical consultation and the instituting of the discourse of medicine in its place, the progressive disintegration of a sense of self and the amplification of the medical ontology with respect to the presence or absence of a physiological cause.

Scarry describes her use of the term 'language-destroying' to mean the destruction of the 'power of verbal objectification, a major source of our self-extension, a vehicle through which the pain could be lifted out into the world and eliminated' (1985: 54).[11] In this sense, 'language' refers to the social construction of subjectivity, of self in relation to others within culture; the creation of distinctions through language: the making of the (meaning of the) world (genesis). But, like Foucault, Scarry ignores the significance of the embodied subject being importantly a *sexed* embodied subject. The fact that the speaking subject is gendered male or female is integrally related to the 'problem' of the unrepresentable. Grosz refers to Kristeva's notion of the 'chora', the receptacle that constitutes the debt representation owes to what it cannot name or represent, which she (Kristeva) characterises in terms of its maternal and spatial origins (Grosz, 1995: 84). But Grosz is concerned not to render that-which-is-deemed-unrepresentable as inherently mysterious, as undecipherably enigmatic. She writes, 'there is no object of reference, no concept or term so inherently opaque that it cannot be known' (1985: 223). However, she resists the position that articulateness and representation 'are in themselves a virtue' (1985: 222).

The important point here is how 'language' is understood. When language is construed as a representational system that empowers the speaker to bring objects into existence, make them real through naming them, then it follows that the logic of identity/difference will ensure the privileged status of he (or she) who succeeds in identifying 'it' (something, anything) and indeed, being identified *by* language *as* 'it'. Language becomes, as Braidotti (1994: 44) states (with reference to Baudrillard), the ultimate prosthesis. A critique of this (dominant) conceptualisation of language proposes the articulation of speech as a process of symbolic exchange[12] that precisely creates subjects of social life.[13] Rather than establishing or representing what 'is' in essentialist terms, language evokes, points towards, or 'seduces' (in Baudrillard's terms, 1979 trans 1990).

The expressed motive of torture was analysed by Scarry to be a fictitious motive. The expressed motive of medicine is to help, to heal, to relieve suffering. I don't want to suggest a parallel here and argue that this is also fictititous, but I do want to explore the possibility that this singular motive overflows its terrain of possibility and serves to obscure an equally significant motive in the medical drama, that of the will to power-knowledge; the will to truth, to be he who knows, he who can say it all, he who can control, make and unmake. When this profoundly gendered 'will' is thwarted, when he whose subjectivity (he who knows) is confronted with she who is in pain, she whose nebulous pelvic interior disrupts his sense of self with that which is or might be unrepresentable, it seems to me that this is possibly one of the most likely sites in our culture for him to project onto her his own fear of that which is unrepresentable, his own 'death drive', to use Kristeva's term (1989) taken directly from Freud, a regressive desire to return to the origin (primarily symbolised by the female womb); a drive that cannot consciously coexist with his identity and therefore has to be projected onto the other, onto the woman. That which is unrepresentable threatens his identity at a supremely deep level, so he projects his fear onto the woman and she becomes the site of that which is unrepresentable; she is forced more and more into invisibility as a subject, forced further and further into the corporeal space as if in the same movement he is positioned progressively further away from it; a rejection of, or even repulsion from, the immanence symbolised by the feminine in de Beauvoir's terms (1949).

The false motive of torture allows, in the context of an appalling abuse of truth, some degree of apparent moral responsibility to accrue to the torturer: the prisoner is revealed as morally irresponsible because he or she betrays his or her family, friends or whatever. In the case of the medical encounter, the motive of helping, the innocence of the appearance of altruism, opens the way for a displacement of responsibility: the responsibility for the inability of the doctor to *actually* help is displaced onto the patient in the form of 'neurosis'. It becomes the patient's fault that she is in pain, she is creating her own pain; the attribute of pain whereby the agency of pain is able to be experienced as the body hurting itself is neatly mapped onto cultural stereotypes of women as 'neurotic', a keyword which *seems* to mean irrational, making a fuss about nothing, concocting a story that is designed to madden and confuse. In fact, the signifier 'neurotic' in this context is curiously devoid of meaning, which precisely mirrors the state of consciousness Scarry talks about in relation to the prisoner's pain, a consciousness that is emptied of content.

Through this mechanism, the pain is denied. The doctor distances himself from the woman's pain and does not acknowledge its existence. The keyword 'neurotic' works in the same direction as the keyword 'betrayal' in the process of torture. And we can say that the construction of the woman as 'neurotic' is in fact a 'betrayal'; it *betrays* the motive of enabling wholeness since the signifier 'neurotic' is the route to the de(con)struction of self and it *betrays* the doctor as he who cannot help, he who plays a role antithetical to the expressed intent of medical practice.

The woman with chronic pelvic pain whom the doctor believes has no 'organic pathology' does not conform to the medical discourse. When the technology of medicine loses its power to display in grand performance the drama of creating 'wholeness', which it attempts to claim as its own achievement, the absence of 'something found' renders the medical agent/actor impotent. The woman returning over and over to the doctor is a voice saying you don't know, you can't help me, your knowledge doesn't work. 'There's nothing wrong with you', with its echo of 'you're neurotic', denies the pain, neutralises the social capability of the woman's voice and functions to further isolate her, for her to experience a further contraction of her space, to open her to becoming increasingly overwhelmed by her body-in-pain, making that pain work to erode her sense of self.

The confidence of the embodied subject diminishes as her self becomes increasingly circumscribed while the voice of the discourse of medical science increases as it replaces the voice of the life-world and the woman's talk is more and more that of medical terminology, physiological explanations and her distress becomes the absence of tissue damage in her pelvic cavity. (Both women with pelvic pain and gynaecologists themselves have told me how women have had laparoscopy after laparoscopy to look and look and try and find something; women fantasise and literally dream of something being removed from their pelvis which relieves the pain.)

Following Scarry's analysis of the role of pain in torture, it could be argued that, in the case of the medical encounter for CPP, the instrumental role of pain is used to create and recreate the power of medicine. Instead of rendering the attributes of pain visible through a weapon and the macabre play of appearance and disappearance, the doctor keeps the pain resolutely invisible through silencing the woman's voice and superimposing his own, the voice of medicine; the doctor uses his agency to force the pain into greater invisibility. And possibly effects a comparable end: the regime of medical power retains the power to determine the realness/the existence of pain and its attributes, through the power-knowledge of its assumed origin.

The medical scene almost appears as a reverse tendency from the appropriation of the attributes of pain into the insignia of power through torture. Rather, the denial of pain and the silencing of the woman-as-subject enforces the double invisibility of pain; it is not 'we who do this to you' but rather 'you who does this to yourself'. But this can also be seen as a point of connection with the sentiment of the torturer: you can solve the problem if you only acknowledge complicity. Scarry commented in relation to the role of torturer 'the most radical act of distancing resides in his disclaiming of the other's hurt' (1985: 57). One way the doctor can potentially exercise power in the medical consultation for CPP could be said to be found in his action of not 'knowing' the woman's pain, an act of disavowal that is also made possible by the metaphysics of medical knowledge.[14]

Confronting Paradoxes

I am interested in exploring the parameters of a feminist approach to the *practice* of understanding chronic pelvic pain and disease that engages the subjectivity of the woman suffering in a way that reconfigures the split between an anatomical body devoid of consciousness on the one hand and an embodied agentic subject on the other; an understanding that confronts this paradox. Its starting point would be a critical epistemology of theorising social phenomena: an epistemology that embraces a feminist critique of positivism, empiricism, absolutism and relativism; an approach to the construction of knowledge that is historically and culturally specific; that is grounded in a realist postulate, acknowledging that the social world is contradictory; an approach, therefore, that brings the role of discourse fully into play.

But such an approach would have to deconstruct, go beyond, the natural science of biomedicine as being incommensurate with the critical epistemology required for an understanding of the social. In metaphysical terms, Western medical and psychological philosophy has exhausted the possibility of somehow attempting to hybridise the natural science conceptualisation of cells, tissues and biochemical processes with the social science of interpretive, non-positivist modes of creating knowledge, to create a so-called interactive paradigm of psychophysiology.

To unravel and try and develop an understanding of the workings of power within the paradigm of Western medicine and how it works to discipline and create experiences of the body (in Foucault's terms) is something that even Foucault himself (1973) claimed to be a daunting prospect. Here, however, I think the crucial point is that it is not a matter of developing a paradigm, or a framework of knowledge. From a feminist perspective we need an approach to the understanding of pain, suffering and disease, that is inextricably linked with practice, that precisely does not fix knowledge within a positivist fiction of the real and its etiology.

If we are not about creating a paradigm, it will not, therefore be a 'theory' that is created to perform an awesome task of universal explanation; it will not be a story told by any one author. Thus, what is of importance is to develop a critical understanding of what is happening when, in Western contexts, women with chronic pelvic pain consult the medical profession and go through a process of loss of self in conjunction with denial of pain.

Do we need a more workable, useful conceptualisation of the notion of psychosomatic illness or psychosomatic pain, or, as Baudrillard (1983) has claimed, will this simply inevitably produce a category that is poorly defined and thus perpetuate the problem of another category that is unrepresentable in medical terms? If the dominance of the visual imperative of empiricist medical science remains unchallenged, then presumably this would indeed be the case.

An article translated from the French written by two male gynaecologists Maruani and Atlan in 1981, on the place of the psychologist in cases of women's pelvic pain, is the only example I have read of an approach to CPP that appears to be grounded in a different set of assumptions and I think raises issues that assist the process of attempting to rethink this whole problematic. Maruani and Atlan foreground the question of the fears of the gynaecologist and focus on the relationship between the doctor and patient as the primary site influencing what happens in terms of the success or otherwise of the medical encounter. They do not talk about *with* or *without* organic pathology and are clear that, in their terms, 'a truly psychosomatic approach' does not privilege one or other site (physiology or psychology) nor envisage a cause–effect relationship between them.

They begin with the woman's voicing of her symptoms as the starting point and discuss the way they perceive that the psychology of the interaction has, potentially, the elements of a sado-masochistic dynamic (along the lines of 'I am hurting, you have the power to stop it and you can't or won't, watch me hurt more and more'). Maruani and Atlan state that the gynaecologist is an actor, a subject and not a spectator. The symptoms are not passively 'observed' but are, at least in part, created in the context of the interaction. They suggest that the gynaecologist does not observe a fact in itself, but a fact 'for them' (and the 'for them' means both from their viewpoint and also the sense of given to them by the patient).

Tentatively, what I think we can take from Maruani and Atlan is that until the subjectivity of the actors – both doctor and woman patient – is critically engaged in the medical encounter, we will not make progress in challenging the dualistic and fixed paradigm of medicine, nor indeed the power relations that are, at least in part, generated by it. A critique of the physiological/psychological dichotomy implicit in this paradigm opens onto a representation of the body that might better be characterised as an 'open materiality', in Grosz's terms (1994), as a discursive field that is simultaneously a surface inscribed with meaning and an interior where meanings are made visceral.

Notes

1. For a more in-depth discussion of the medical, psychiatric and psychological literature on chronic pelvic pain see Grace (1997).
2. It is important to recognise that this is only one of many names given to this 'condition' but the one, for my purposes, that is the most revealing.
3. Lacan called it 'ego psychology' (see Benvenuto and Kennedy, 1986), to refer to the distorted assumption that psychoanalysis is concerned to divest the individual of neurotic, conflictual and contradictory emotions in an attempt to build a unified, non-conflicted, fully functional ego. According to Lacan's own theory, such an ego is precisely an illusion and psychoanalysis is a process, a practice that assists the analysand to come to terms with this reality through understanding.
4. I am using 'dominant' and 'subordinate' term after Gross (1986).
5. 'Biopower' is Foucault's term for an era characterised by technologies of power that have as their modus operandi the management, control and surveillance of the production of life, by contrast with a previous era where power was exercised in systemic terms through the control of death and systems of punishment.
6. I am using 'him'/'he' to indicate a masculinist position rather than wishing to refer to the sex of the person.
7. Chronic pelvic pain with no obvious organic pathology has been often claimed to be one of the most 'perplexing problems' facing gynaecologic medicine.

8. Although this terminology is sexist, it is not essentialist; in postulating the emergence of man Foucault allows the envisioning of his end. The prototype of the humanist subject was certainly male.
9. It is noteworthy that Scarry does not refer once to Foucault's analysis of the historical transition from one form of power to another in Europe and its significance regarding the end of precisely the public display of torture to institute power and control bodies (Foucault, 1975, trans 1977).
10. Scarry's use of the concept 'political instability' is best understood in the sense of ethical and normative instability rather than in the historical sense of threats to political rule; consistent with Foucault's formulation, there is no power without resistance, no absolute and stable power. Thanks to Livia Käthe Wittmann and Renee Heberle for valuable discussion on this point.
11. In the context of torture, Scarry is presumably referring to the capacity of language to enforce a recognition of the subjecthood of the human agent and the subsequent impossibility of the continued infliction of pain.
12. 'Symbolic' is used here in Baudrillard's formulation where the symbolic is precisely that which can never be represented to reveal its 'truth' or essence. The symbolic is always ambiguous.
13. Joyce McDougall's work on a psychoanalytic approach to psychosomatic illness (1989) provides interesting observations and insights when considered in this context alongside Scarry's analysis. Where Scarry claims that extreme bodily pain is language destroying and in this sense evocative of pre-Oedipal experience (the period in the infant's life prior to the formation of language), McDougall is of the view that psychosomatic conditions are themselves precisely not of the order of neuroses but are better characterised and understood in pre-Oedipal terms, prior to language, in other words formed at a time when the individual can only express pain through the body.
14. In a recent work Scarry (1994) presents a compelling analysis of the way pharmaceutical companies design their advertisements for analgesics. She argues that when these advertisements are aimed at doctors, their primary intent is to represent the internal reality of the sensation of pain which is contrasted to those advertisements aimed at those suffering pain where this intent is precisely absent.

Acknowledgements

My appreciation and very grateful thanks to Livia Käthe Wittmann for her helpful and insightful reading of drafts of this chapter. Thanks also to Louis Arnoux and Chris Prentice for their comments on a draft of this text.

References

Baudrillard, J. (1979, trans. 1990), *Seduction,* London: Macmillan.

Baudrillard, J. (1983, trans. 1990), *Fatal Strategies,* London: Pluto Press.

Bekker, F., Peters, L. and Bayer-Rots, M. (1990), 'Chronic Pelvic Pain Patients and Conflicts in Gender Identity', in *Journal of Psychosomatic Obstetrics and Gynaecology*, vol. 11, no. 1, pp. 19–28.

Benvenuto, B. and Kennedy, R. (1986), *The Works of Jaques Lacan – An Introduction,* London: Free Association Books.

Braidotti, R. (1994), *Nomadic Subjects. Embodiment and Sexual Difference in Contemporary Feminist Theory,* New York: Columbia University Press.

Castelnuovo-Tedesco, P. and Krout, B.M. (1970), 'Psychosomatic Aspects of Chronic Pelvic Pain', in *Psychiatry in Medicine*, vol. 1, no. 2, pp. 109–126.

de Beauvoir, S. (1949), *The Second Sex*, London: Pan Books.

Fisher, S. and Todd, A.D. (1983), *The Social Organisation of Doctor-Patient Communication*, Washington: Centre for Applied Linguistics.

Foucault, M. (1966, trans 1970), *The Order of Things. An Archaeology of the Human Sciences*, New York: Random House.

Foucault, M. (1973), *The Birth of the Clinic. An Archaeology of Medical Perception*, New York: Random House.

Foucault, M. (1975, trans 1977), *Discipline and Punish. The Birth of the Prison*, New York: Vintage Books.

Friederich, M.A. (1976), 'Psychological Aspects of Chronic Pelvic Pain', in *Clinical Obstetrics and Gynaecology*, vol. 19, pp. 399–406.

Gidro-Frank, L., Gordon, T. and Taylor, H.C. (1960), 'Pelvic Pain and Female Identity – A Survey of Emotional Factors in 40 Patients', in *American Journal of Obstetrics and Gynecology*, vol. 79, no. 6, pp. 1184–1202.

Grace, V.M. (1995a), 'Problems Women Patients Experience in the Medical Encounter for Chronic Pelvic Pain: a New Zealand Study', in *Health Care for Women International*, vol. 16, no. 6, pp. 509–519.

Grace, V.M. (1995b), 'Problems of Communication, Diagnosis and Treatment in the Use of the New Zealand Health Services for Chronic Pelvic Pain: a Quantitative Analysis', in *Health Care for Women International*, vol. 16, no. 6, pp. 521–535.

Grace, V.M. (1997), 'The Mind/Body Dualism in Medicine: the Case of Chronic Pelvic Pain. A Critical Review of the Literature', in *International Journal of Health Services* (forthcoming).

Gross, L. (1986), 'Derrida and the Limits of Philosophy', in *Thesis 11*, vol. 14, pp. 26–43.

Gross, R.J., Doerr, H., Caldirola, D., Guzinski, G.M. and Ripley, H.S. (1980–1981), 'Borderline Syndrome and Incest in Chronic Pelvic Pain Patients', in *International Journal of Psychiatry in Medicine*, vol. 10, no. 1, pp. 79–96.

Grosz, E. (1994), *Volatile Bodies. Towards a Corporeal Feminism*, Bloomington: Indiana University Press.

Grosz, E. (1995), *Space, Time, Perversion. The Politics of Bodies*, New York: Routledge.

Habermas, J. (1987), *The Theory of Communicative Action. Vol. 2. Lifeworld and System*, Boston: Beacon.

Hodgkiss, A.D. and Watson, J.P. (1994), 'Psychiatric Morbidity and Illness Behaviour in Women with Chronic Pelvic Pain', in *Journal of Psychosomatic Research*, vol. 38, no. 1, pp. 3–9.

Kristeva, J. (1989), *Black Sun: Depression and Melancholia*, New York: Columbia University Press.

Maruani, G. and Atlan, P. (1981), 'Algies Pelviennes Chroniques: la Place des Psy' ('Chronic Pelvic Pain: the Place of the Psychologist'), *Genitif*, vol. 3, no. 9, pp. 3–10.

McDougall, J. (1989), *Theatres of the Body – A Psychoanalytical Approach to Psychosomatic Illness*, London: Free Association Books.

Mishler, E.G. (1984), *The Discourse of Medicine*, Norwood, N.J: Ablex.

Scarry, E. (1985), *The Body in Pain. The Making and Unmaking of the World*, New York: Oxford University Press.

Scarry, E. (1994), *Resisting Representation*, New York: Oxford University Press.

Todd, A.D. (1989), *Intimate Adversaries. Cultural Conflict between Doctors and Women Patients*, Philadelphia: University of Pennsylvania Press.

The Embodied Experience of Largeness

A Feminist Exploration

Jenny Carryer

Women in Western societies are experiencing a period where the desirable body shape for a woman is particularly lean. Walden (1985) noted that whilst size consciousness became quite focused at the turn of the century, requirements for women to have a slender body had become especially pronounced within the 20 years prior to 1985. In the last 10 years this focus on slenderness has escalated unabated. Simultaneously, the role of the media in proliferating normative images and discourses has intensified, producing limitless opportunities for images of the thin female body to be represented as an ideal.

Slenderness is no longer viewed as an arbitrary genetic trait (Atrens, 1988), occurring serendipitously in Western women. Rather it is viewed as an essential even defining aspect of beauty and linked to discipline and vigilance. Caputi (1983) notes that whereas thinness signifies beauty, health, class, grace, discipline and goodness; fat is equated with ugliness, stupidity, illness, sloth, self-indulgence and moral weakness. Many women consider their size to fall outside the norm and are experiencing, and sometimes articulating, their distress. This chapter will focus specifically on the experience of women whose bodies are already large despite their best efforts to resist weight gain.

The discussion arises from research in progress, which has involved a period of co-structured interviews with nine women whose large body size falls well outside accepted norms. Feminist methodological assumptions underpin the research and the interview technique has been highly reflexive. This meant that throughout the interviews I shared the insights of my ongoing reading and thinking and many aspects of my personal experience as a large woman. Participants were continually encouraged to direct the focus of enquiry, although I asked many specific questions as I followed issues raised by other participants or pursued areas that seemed pertinent to me. This process has supported participants and assisted them to feel more comfortable talking about a life experience which is more usually private and often extremely painful.

In this chapter I will present some of the women's experiences as they have shared them with me. In order to describe and contextualise the women's experiences, I will first consider the impact of medicalisation and then consider issues of gender and femininity as they impact on women's embodied largeness.

Medicalisation

Medicalisation refers to a process whereby Western medicine turns its gaze toward aspects of women's lives and bodies, renders them problematic and focuses attention on treatment to achieve a cure. The current focus on normal and abnormal body sizes can be read as involving a significant degree of medicalisation. Ever since slimness became an integral part of current concepts of beauty, medicine, according to Walden (1985), has taken a deep and sustained interest in so-called obesity. The vast medical literature about obesity reveals a degree of contradictory information. As a nurse working closely with other nurses and medical practitioners, I had always been certain that any excess body weight was detrimental to health. A reading of the medical research literature has shown, however, that there is information which significantly modifies such conventional medical wisdom and yet this critique has remained marginal within what is an apparently neutral scientific endeavour. Poorer health outcomes, especially for women, cannot be conclusively demonstrated over a much wider range of weight levels than is popularly understood to be desirable. Yet medicine does not intervene in the public, non-medical discourse which suggests that striving for a very low body weight is meritorious.

I will attempt to describe some of the apparent contradictions in medical teaching about 'obesity'. Walden suggests that:

> Medicine turns to what has been judged to be naturally ugly or deviant and then develops aetiological accounts in order to explain and treat in a coherent fashion a manifold of displeasing signs and symptoms.
>
> (Walden, 1985: 367)

The women participants in my study, by virtue of having larger than average bodies, are deemed to suffer from a disease syndrome known as obesity. Obesity as a 'disease' appears to have no absolutely proven cause, no completely guaranteed cure, conflicting consequences and is diagnosed almost entirely on visual grounds alone. The medical focus on the pathology of size when considered alone and without the benefit of political analysis appears to be a simple concern for disease prevention based on the premise that all weight over a particular level is detrimental to health and longevity. A feminist analysis allows us to see this situation in different ways, but first it is useful to examine the discourse of Western medicine itself, as it relates to body size.

The Discourse of Medicine

Obesity is regarded in medical terms as a serious, prevalent and refractory disorder (Brownell and Wadden, 1992). It is believed to be associated with increased risk of hypertension, diabetes, cardiovascular disease and other illness (Bray, 1986; Pi-Sunyer, 1991). Swinburn (1995), the medical director of the National Heart Foundation, has commented that there is no doubt that there is a major epidemic of obesity affecting all developed countries and many developing countries. He notes that this epidemic has been described as a tidal wave. Meisler and St Jeor (1996) note that currently the healthy weight targets are exceeded by one in three adults.

Van Itallie (1979) recorded that the prevalence of obesity had doubled since 1900 which notably is, according to Walden (1985), about the time concern with corpulence became a feature of social existence. According to Marston and Raper (1987), caloric intake has been stable or declining since that time, but they also offer evidence for declining physical activity in normal daily lifestyles and work requirements. A Consensus Development Conference held in the United States of America in 1985 was unequivocal in its conclusion that fatness is dangerous to health and highly undesirable. Ernsberger and Haskew (1987), in an overview of current medical wisdom and in response to the consensus conference report, observe that medicine consistently argues for a relationship between increased body size and poor health outcomes.

The past 40 years has seen considerable debate and research into the cause of obesity. Brownell and Wadden (1992) in an article reviewing that research, note that much of it has looked for a single cause to explain a single entity. Over many years theories have ranged from psychoanalytic (Stunkard, 1988) behavioural (Stuart, 1967) and genetic (Stunkard, Harris, Pedersen and McClearn, 1990) all cited in Brownell and Wadden (1992). Research has looked both at the behaviour of overweight people and the physiological status of the human body in order to understand variables such as the resting metabolic rate, fat cell numbers and genetic components of weight variation.

Despite no real consensus or understanding of the cause or consequences of obesity, medicine has, for a long time, sanctioned a number of treatments for large body size, the most common and enduring being reduction dieting. Various pharmaceutical measures have been tried including appetite suppressing drugs and especially amphetamines (Hawkins and Sprague, 1986 cited in Ernsberger and Haskew, 1987). Pharmaceutical measures included the use of the rainbow pill, which combined an amphetamine (central nervous system stimulant) a diuretic (to remove excess fluid) and thyroxine (to speed the metabolism). This pill was in vogue from the 1940s through to the early 1970s. Ernsberger and Haskew (1987) note the use of jaw wiring since 1974. This is the process of wiring the teeth together to permit only the passage of fluids for a period of months. They also note the annual performance of 30,000 gastric bypass operations in the United States by 1980 increasing to 50,000 by 1982. They note that, in the quest for a smaller body, people have endured these treatments often with considerable suffering and some degree of danger or risk.

Brownell and Wadden (1992) discuss the repeated reduction dieting which characterises the lives of many large people (especially women) and acknowledge that it does not produce a lasting solution. They note that:

> ... [o]bese persons may bear a legacy of shame and failure from cycles of weight loss and regain. Our patients have undertaken an average of five major diets on which they lost and eventually

> regained a total of 56kgs. Despite their efforts, they have watched their weight increase from 70kgs, at 17 to 100kgs. at 40.
> (Brownell and Wadden, 1992: 513)

Generally the overt pronouncements of medicine, with regard to obesity, are accepted without challenge by the community at large and usually by practitioners of medicine and nursing as well. I am mindful here of Lorraine Code's comment about what she calls 'the science has proved rhetoric' (Code, 1995: 27) which allows a degree of blind faith in any judgement which purports to arise from the pristine objectivity of positivist science .

Ernsberger and Haskew (1987) suggest that the automatic linking of large body size with poor health needs considerable modification. In a comprehensive review of significant research they argue that the 1985 American Consensus Statement on the health implications of obesity represented an unbalanced view of adiposity and health. In their research review they note that although medical literature has documented elevated risk factors in heavy people, these risk factors fail to translate into high mortality rates. Ernsberger and Haskew note, surprisingly, that adiposity protects against many diseases and is associated with a more favourable outcome in others. They note that many studies show that maximum longevity is associated, in women at least, with above average weight.

Bodroff and Bjorntorp (1991) further suggest that the female pattern of fat distribution with heavy hips thighs and buttocks (pear shaped) is associated with virtually no impairment of health except possible joint damage. The male pattern of small hips and thighs but rotund abdomen (apple shaped) is thought to be more closely associated with elevated risks of many of the degenerative diseases.

Medical literature on the relationship between weight and health has also been criticised by Rothblum (cited in Brown and Rothblum, 1989) who notes two major design flaws. In studies relating health status to weight levels there has not been control for the variable of history of reduction dieting, possibly because it is difficult to find a large woman who has never dieted. Further, it is constantly reported that the prevalence of obesity is much greater in lower socioeconomic groups (Sobal and Stunkard, 1989), yet this variable is also not routinely controlled. It is well known that health status is related to socioeconomic status, with low socioeconomic groups having the poorest health and higher incidences of many disorders (Nelson, 1994).

There is, therefore, some degree of discrepancy within the discourse of medicine between knowledge available from research and the knowledge which reaches the public domain. It has been my experience in working with, speaking to or teaching medical practitioners and nurses that they are generally unaware that there is at the very least potential challenge to their belief that any degree of obesity automatically constitutes a major health risk. This belief is not usually publicly modified with reference to gender difference and the variables of fat distribution and their significance. It is also not publicly modified in respect of degrees of obesity, or the relationship between weight gain and age increase. In general, medical and nursing personnel are supportive of the practice of reduction dieting despite considerable research-based evidence that it is ineffective as a long-term or permanent solution for large body size (Bennet and Gurin, 1982; Ernsberger and Haskew, 1987; Atrens, 1988; Cowen, Clifford, Williams, Walsh and Fairburn, 1995).

Stunkard *et al.* (1986) noted that there was a demonstable relationship across the whole range of body fatness between adoptees and birth parents. Swinburn (1995) says that heritability of 'obesity' is about 33 per cent. Generally such information is not public knowledge, along with the necessary recognition that many people are not naturally thin and will always be somewhat larger than others despite having healthy eating habits. There is a growing awareness in the popular press that reduction dieting is not a permanent solution but there is no evidence to suggest that reduction dieting has decreased. It is evident that some information remains as a marginal discourse while other information carries the weight of dominant discourse firmly embedded in institutions such as medicine.

It is not well known that out of more than 30 international studies of the eating habits of large people, only one study demonstrated overeating on the part of large subjects (Bennet and Gurin, 1982). Despite this, reduction dieting remains a multi-billion dollar enterprise. Swinburn (1995) reports the cost of the diet industry in the United States as 33 billion dollars annually. In many instances the diet industry is publicly endorsed by medical professionals holding positions of authority. The diet industry capitalises on both women's pain and the medical approval that reduction dieting receives. The net result for the women in this study is that they assume that they must overeat because it is such a taken-for-granted assumption and they feel that reduction dieting is the only option available to them.

Medical approval of small body size at any cost is just one aspect of the pressure to be small experienced by many women and most intensely by large women. Socio-cultural and political factors are equally if not overarchingly significant contributors to the endless quest for smallness.

Body, Gender, Femininity

The quest for a slender body is clearly linked to current constructions of femininity. Women's anatomy (morphology) has become at one with her destiny and few parts of women's bodies are not considered open to improvement. The approved and required slender body for women occurs effortlessly for some but is virtually unobtainable for many and there is evidence that the goal has become increasingly remote (Kissling, 1991). Kissling notes that current cultural icons of beauty and thinness are more and more remote from the reality of women's bodies creating a vast gulf between achievement and despair. These emotions are addressed by increasing levels of self-surveillance and efforts toward self-normalisation. Spitzack (1990) carried out research into women's body talk and noted, amongst other issues, that virtually all women believe that their lives would be enhanced by the loss of at least 14 or more pounds. In the process of dieting to lose that 14 pounds, the relationship to food may be irrevocably altered (Polivy and Herman, 1985).

Evidence for the detrimental consequences and apparent ineffectiveness of reduction dieting is growing (Bennet and Gurin, 1982; Polivy and Herman, 1985; Ogden, 1992). The women in my study had experienced a number of reduction diets which, due to their determination and vigilance, were, in the short term, successful. As Atrens (1988) describes, the body responds to the sensed starvation of a reduction diet and responds by conserving fat and slowing metabolism. More recently Cowen, Clifford, Williams, Walsh and

Fairburn (1995) have demonstrated that even moderate dieting alters the activity of a brain transmitter involved in regulating appetite and food intake. Hence women may begin to diet at a relatively low and quite healthy weight and simply cause a steady escalation of their weight. Failure intensifies their self-criticism and disgust and they usually respond by vowing to succeed in the next diet.

The attainment and maintenance of a slender body is an integral part of what is assumed to be appropriate feminine appearance. According to Bartky (1990), the disciplinary practices in which women engage are part of the 'process by which the ideal body of femininity – and hence the feminine body-subject – is constructed; in doing this, they produce a "practised and subjected body", i.e. a body on which an inferior status has been inscribed'. (Bartky, 1990: 71). The battle against the burgeoning body is aptly described within this context and participants in the research in progress can be seen to be engaged in a project which is protracted and never complete.

Bordo (1989) suggests that the bodily inscription of femininity is especially problematic as current cultural norms for femininity have become increasingly contradictory and conflicting. Women remain positioned as the primary feeders and nurturers of others especially in the heterosexual context. Yet the requirement for a slender body makes the feeding and nurturing of women's own bodies problematic. There is an internal contradiction in the association of giving food and love to others while exercising restraint, control and restriction toward their own needs.

Chernin (1981) addresses another aspect of the contradiction in suggesting that femininity is both feared and revered and simultaneously sought after and denied. She suggests that struggling to embody and enact a femininity which is emphasised yet also devalued, creates a double bind and an essentially unwinnable situation for women. Bordo (1989) argues that the pursuit of slenderness offers the chimera of meeting, through the body, these essentially contradictory demands of the contemporary construction of femininity. By this I understand Bordo to mean that the process of losing weight embodies masculine ideals of control, discipline and strength whilst concurrently and paradoxically supporting a movement towards a goal of smallness, delicacy and presumed attractiveness.

It is notable that the reduction of women's body fat is in itself contradictory. It represents a move towards femininity in the attainment of smallness and fragility but at the same time increasing slenderness eradicates those parts of the body which are clearly female. Caskey (1986: 176) illuminates this by considering the relation of body fat to femininity. She suggests that 'from sexual maturation to the end of life, body fat, by its relation to female hormones is essentially resistant to the conscious wishes of its owner'. The parts of the female body which are clearly different from men's bodies are those where fat accumulates, i.e. the breasts and hips. As women gain weight, their female shape is more and more defined; as they become smaller, the differences are less exaggerated. This is a strange and contradictory bind within which women live out their daily existence.

Bordo (1990) argues that slenderness functions within a modern normalising machinery of power in general and it especially functions to reproduce gender relations. She suggests that preoccupation with fat, diet and slenderness functions:

> ... [a]s one of the most powerful 'normalising strategies' of our century, ensuring the production of self-monitoring and self-disciplining docile bodies sensitive to any departure from social norms and habituated to self-improvement and transformation in the service of those norms.
>
> (Bordo, 1990: 85)

Participants in this research project have spent their lives quite literally battling their unruly bodies; bodies which stubbornly resist a much sought after conformity. Their stories are almost all strongly suggestive of a natural tendency to gain weight. Pregnancies have been especially difficult. They have frequently been vigorous and determined dieters but despite that, in fact perhaps because of that, have watched their bodies grow ever larger. In the following section I will share some commonalities in the women's experiences as they have described them.

Their childhood, almost always as the daughters of large mothers and sometimes fathers and other close relatives, is characterised by an emerging sense of difference and a growing awareness of being 'bonny' or 'chubby' or 'solid'. Frequently there is a moment in time in which their self-concept is suddenly catapulted into seeing themselves as fat or 'too big'. For Clare (aged 15) it was the time the sewing teacher tweaked the waist band of the skirt she was making and mentioned that fewer potatoes would be a good idea. For Elizabeth at the age of 13 it was dancing away the night with a boy at a third form dance and seeing to her horror in photos afterwards that she had physically dwarfed him. Celia remembers not eating for four days prior to starting at university so that she could arrive looking suitably 'pale and interesting'. This seemed more in keeping with the current images portrayed in the *Seventeen* magazine which did not include her own slightly sturdy and rosy-cheeked appearance. For other participants inability to run as fast or not looking good in the right clothes engendered an emerging shame. In many cases their mothers' frequent dieting helped make connections about appropriate management of a too large body and if not, then every woman's magazine suggested multiple variations of the same theme. Start a diet today and let the new you emerge.

All participants have had considerable engagement with reduction dieting. Only the two youngest participants could not reel off a list of temporarily successful diets. At the least they have always had a constant level of food intake awareness, waking in the morning and thinking 'today I will diet' or feeling guilty about the previous day's imagined excesses. Many of the participants have lost and regained over 50kgs in the course of several serious and protracted reduction diets under the auspices of 'Weight Watchers', 'Jenny Craig' or 'Diet Clinic' or other variations. Those who have not utilised formal dieting venues have used informal methods with the same varying degrees of short-term success. These women have demonstrated enormous determination in enduring prolonged food deprivation requiring constant planning, special shopping, individual food preparation and considerable difficulty in social settings. Simultaneously, embodied largeness has been a persistent feature of their daily existence.

Celia described tipping a cup of scalding coffee over her thighs at an outdoor party. She sat and quietly burned because she knew if she revealed what had happened she would be offered help necessitating body exposure and offers of replacement clothes which would not fit. On the surface Celia is a confident and articulate professional woman. Clare, as a theatre nurse, has watched with anguish as surgeons contemptuously jiggle the large stomachs of anaesthetised women on the table awaiting surgery. She finds it hard to accept her apparent complicity when she remains a silent participant in the mockery, yet, as a large woman herself, speaking out is too frightening. Celia has a friend who covers her body with flannels as she lies in the bath because she cannot bear the sight of her own body. Maggie describes taking a float plane with a group of friends and learning mid-flight that they would move from the plane to dry land piggybacked by the somewhat diminutive pilot. The rest of the flight and the subsequent disembarkation passed in a blur of humiliation.

Anticipating difficult situations is an essential part of embodied largeness but it is not always possible to successfully anticipate an embarrassing occasion. Withdrawal from social and recreational participation was a common theme for participants who often decided that the threat of ridicule or embarrassment was worse than missing out. There is very little that can be carelessly done without a constant state of vigilance or preparedness that a situation might arise and the body will be exposed to ridicule. To some degree the stigmatisation actively works to impede health by altering the lifestyle of the large woman. As Heather noted:

> You see it seems to me that being overweight is like a trap, because being overweight makes it harder and more painful to exercise, it makes it more difficult to get involved to get out and do things and it makes it more difficult to be taken seriously in terms of health care and yet all of those things – not taking your health care seriously, not enjoying your life and getting out and not exercising – are quite detrimental to health in themselves; it's like a vicious circle that we get trapped in, that all large women get trapped in.

Celia is a highly organised and efficient professional woman who longs to ski and play tennis but withdraws from these activities because she perceives her body size to be inappropriate for such pursuits. She experiences contradictions between her own self-image and interpretations of her real body image. She says:

> ... I don't do a lot of physical things and then when I do, I think of myself as a really active energetic person and then I look at myself objectively and I see a rather lumpy looking person who looks like she couldn't organise her way out of paper bag ...

Women have historically endured such bodily restrictions as corsets, lacing, bustles, hoops and shoe heels to accentuate or diminish the necessary body parts. Currently the colonisation of the female body (Smith, 1989) has gone a good deal further in that it now requires that the flesh itself be disciplined into shape and the lives and activities of the woman similarly constrained.

Consciousness of being 'too fat' intrudes very deeply into these women's lives and this was epitomised by a comment made by Elizabeth, who said:

> I even think to myself, this is really horrible and I hope it doesn't happen, but I think if something happens to me, if I ended up with cancer at least when I died I'd be thin and when they carry out the coffin they're not going to be all straining and puffing and blowing. ... I've actually said to my husband, 'Oh well, if I did get the big C at least I'd be thin and then you would be pleased'.

This comment from Elizabeth neatly endorses the research-based comment by Bovey (1989) that a large woman can be seriously ill but frequently receive comments from well-meaning friends to the effect that she shouldn't worry too much because at least she is losing some weight. Such comments give large women very clear messages about the relative worth of their health and wellbeing as opposed to their attainment of a suitably small body.

In the absence of the dubious blessing of serious illness, women try to achieve the desired goal of slenderness by reduction dieting. But the dieting woman according to Spitzack (1988) moves into a no-win situation. Spitzack notes that in the unsatisfactory process of reduction dieting, a woman must simultaneously transcend her bodily need for sustenance while still living and attempting to function in her body. Hence the proliferation of diet slogans such as 'mind over matter', 'you are what you eat' and 'it is just will power'. The female body becomes an enemy whose excesses the mind must conquer. Women are asked to learn to dislike and reject the very body they must inhabit and to devote considerable energy to pursuing an increasingly unattainable level of slenderness.

The Western cultural identity available to women is one which implicitly leaves no room for the large-bodied woman. She is left with the need for constant action, constant vigilance and the possibility of only temporary success unless she never relaxes. The rich diversity of female shape, the association of body fat with women's sexuality and the natural increase in weight with age has been rendered inherently problematic. Spitzack (1988) asks 'what better way to keep women in a state of subordination than to convince them through their own actions that they are failures, even in managing their own bodies?' (Spitzack, 1988: 70).

For the women in the current study there are no simple solutions and limited choices. Continuing to diet offers a sense of hope but one which is frequently defeated. Ceasing to diet and accepting a larger body means living outside the bounds of acceptability and enduring a degree of stigmatisation in very many settings. Moving backwards and forwards between dieting and not dieting is a frequent outcome and one which supports poor eating habits and inevitable continuing weight gain.

The experience of embodied largeness is one of pain. The pain arises from inhabiting a body which is socially constructed as both pathologised and stigmatised. In three years of talking with the nine participants in my research project, I have searched intently for signs of their resistance to being positioned as passive victims of a dominant framing of the meaning of their body size and shape. On one level I can see a participant who is using her considerable talent to write a work of fiction about embodied largeness; another has gone to work in an up-market shop selling designer clothes for large women. Yet another participant utilises every commercial trapping of feminine beauty and works as a dispenser of beauty advice and products. These woman have positioned themselves in locations of resistance just as I have in the very act of carrying out and writing about this research.

Most of us have, with trepidation but growing conviction, relinquished the dieting and regaining cycle in favour of rediscovering the use of healthy food for the simple sustenance of hunger. Accordingly we have relinquished the one apparent hope for slenderness. Within feminist writing there has been a tendency to talk of working toward acceptance of fatness. At the point of nearing completion of this research project I can only argue that for participants in this project 'fat acceptance' seems a remote dream.

References

Atrens, D. (1988), *Don't Diet,* Australia: Schwartz, Corgi and Bantam.

Bartky, S.L. (1990), *Femininity and Domination: Studies in the Phenomenology of Oppression,* U.S.A.: Routledge.

Bennet, W. and Gurin, J. (1982), *The Dieters' Dilemma: The Scientific Case Against Dieting as a Means of Weight Control,* N.Y.: Basic Books.

Bodroff, B. and Bjorntorp, P. (eds) (1991), *Obesity,* Philadelphia: J.P. Lippincott.

Bordo, S. (1989), 'The Body and the Reproduction of Feminity: A Feminist Apropriation of Foucault', in Jaggar, A.M. and Bordo, S.R. (eds), *Gender/Body/Knowledge*, London: Rutgers University Press, pp. 13–33.

Bordo, S. (1990), 'Reading the Slender Body', in Jacobus, M., Keller, E.F. and Shuttleworth, S. (eds), *Body/Politics*, N.Y.: Routledge, pp. 83–112.

Bovey, S. (1989), *Being Fat is Not a Sin,* U.K.: Routledge.

Bray, G.A. (1986), 'Effects of Obesity on Happiness and Health', in Brownell, K.D. and Foreyt, J.P. (eds), *Handbook of Eating Disorders: Physiology, Psychology and Treatment of Obesity, Anorexia and Bulimia,* New York: Basic Books.

Brown, L.S. and Rothblum, E.D. (eds) (1989), *Fat Oppression and Psychotherapy: A Feminist Perspective,* N.Y.: Haworth Press.

Brownell, K.D. and Wadden, T.A. (1992), 'Etiology and Treatment of Obesity: Understanding a Serious, Prevalent and Refractory Disorder', in *Journal of Consulting and Clinical Psychology*, vol. 60, no. 4, pp. 505–517.

Caskey, N. (1986), 'Interpreting Anorexia Nervosa', in Suleiman, S.R. (ed.), *The Female Body in Western Culture*, Cambridge, MA.: Harvard University Press, pp. 174–194.

Caputi, J. (1983), 'One Size Does Not Fit All: Being Beautiful, Thin and Female in America', in Geist, C. and Nachbar, J. (eds), *The Popular Culture Reader,* Bowling Green Ohio: Bowling Green University Press, pp. 186–204.

Chernin, K. (1981), *The Obsession: Reflections on a Tyranny of Slenderness,* New York: Harper and Row.

Code, L. (1995), *Rhetorical Spaces: Essays on Gendered Locations,* New York: Routledge.

Cowen, P.J., Clifford, E.M., Williams, C., Walsh, A.E.S. and Fairburn, D. (1995), 'Why is Dieting so Difficult?', in Scientific Correspondence, *Nature*, vol. 376, p. 557.

Ernsberger, P. and Haskew, P. (1987), 'Monograph Edition: Rethinking Obesity, an Alternative View of its Health Implications', in *The Journal of Obesity and Weight Regulation*, vol. 6, no. 2, pp. 3(59)–81(137).

Kissling, E.A. (1991), 'One Size Does Not Fit All, or How I Learned to Stop Dieting and Love the Body', in *Quest*, vol. 43, pp. 135–147.

Kuczmarski, R.J. (1992), 'Prevalence of Overweight and Weight Gain in the United States', in *American Journal of Clinical Nutrition*, vol. 55 (Suppl), pp. 495S–502S.

Meisler, J.G. and St Jeor, S. (1996), 'Summary and Recommendations from the American Health Foundation's Expert Panel on Healthy Weight', in *American Journal of Clinical Nutrition*, vol. 63 (Suppl), pp. 474S–477S.

Marston, R. and Raper, N. (1987), 'Nutrient Content of the U.S. Food Supply', in *National Food Review*, Winter-Spring, NFR, vol. 36, pp. 1392–1398.

Nelson, M.A. (1994), 'Economic Impoverishment as a Health Risk: Methodologic and Conceptual Issues', in *Advances in Nursing Science*, vol. 16, no. 3, pp. 1–12.

Ogden, J. (1992), *Fat Chance*, U.K.: Routledge.

Pi Sunyer, F.X. (1991), 'Health Implications of Obesity', in *American Journal of Clinical Nutrition*, vol. 53, pp. 1595S–1603S.

Polivy, J. and Herman, C.P. (1985), 'Dieting and Binging: A Causal Analysis', in *American Psychologist*, vol. 40, pp. 193–201.

Smith, H. (1989), 'Creating a Politics of Appearance', in *Trouble and Strife*, vol. 16, Summer, pp. 36–41.

Sobal, J. and Stunkard, A.J. (1989), 'Socioeconomic Status and Obesity: A Review of the Literature, in *Psychological Bulletin*, vol. 105, pp. 260–275.

Spitzack, C. (1988b), 'Body Talk: the Politics of Weight Loss and Female Identity', in Bate, B. and Taylor, A. (eds), *Women Communicating: Studies of Women's Talk*, Norwood, N.J.: Ablex Press, pp. 51–54.

Spitzack, C. (1990), *Confessing Excess*, N.Y.: State University Press.

Stunkard, A.J., Sorenson, T.I.A., Harris, C., Teasdale, T.W., Chakraborty, R., Schull, W.J. and Schulsinger, F. (1986), 'An Adoption Study of Human Obesity', in *New England Journal of Medicine*, vol. 314, pp. 193–198.

Swinburn, B.A. (1995), 'Origins and Consequences of Obesity', *Report on the CIBA Foundation International Symposium No. 201*, held in Kingston, Jamaica, 27–30 November.

Van Itallie, T.B. (1979), 'Obesity: The American Disease', in *Food Technology*, pp. 43–47.

Walden, K. (1985), 'The Road to Fat City: An Interpretation of the Development of Weight Consciousness in Western Society', in *Historical Reflections*, vol. 2, no. 3, pp. 331–373.

8

'Maiden Diseases' and the Pathologising of Female Puberty and Adolescence

Marion de Ras

According to medical, psychological, sociological, popular and feminist literature, the Western world over the last two to three decades has experienced an epidemic of eating disorders, notably anorexia nervosa, known as the 'hunger disease' and, more recently, bulimia nervosa, known as the 'binge and purge disease'.[1]

In general it is assumed that these illnesses are:

(a) *modern* illnesses which are foremost bound to female puberty and adolescence
(b) *psychosomatic* illnesses which can be explained through a combination of bio-medical, psychological and socio-cultural factors.

Since these illnesses have appeared in the medical registers, doctors, psychiatrists and psychologists, health workers, counsellors, social scientists and feminist therapists have been searching for causes, symptoms, definitions, explanations, treatment and cure of these, as it is assumed, potentially life-threatening conditions.

Somehow this 'epidemic' of eating disorders is paralleled by an 'epidemic' of discourse on the topic, a steady increase of literature and praxis, analysing, problematising, explaining and treating the putative social, psychological and biological conditions of these disorders. However, if it comes to precisely the medical conditions of eating disorders, or more generally 'typically' girls' illnesses, most researchers and theorists must acknowledge that the etiology of these girls' disorders remains obscure.

It is not the etiology *per se* that interests me in this history of young female illnesses. Rather, the obscure etiology forms part of the context for the broader question concerning the process through which certain constitutions of girls and young women became perceived in the course of Western history, as 'female' and belonging to a specific life phase and diagnosed as 'illness'. The assumed incidence of anorexia nervosa and recently bulimia nervosa, reflects more than just the history of an illness; it reflects a history of focusing on and problematising sexual, bodily, social and moral phenomena of girlhood.

My concern is this process of 'focusing on', or 'gazing at', together with what is, in my view, the subsequent product of such focusing: the construction of 'girls' diseases'. To address this construction of female pubertal and adolescent maladies, I will take a closer look at the current explanations of these disorders and question them using a wider socio-historical and cultural framework of Western processes.

Existing Explanations for Girls' Illnesses

What are the current explanations of girls' illnesses, nowadays identified as eating disorders? There are three broad paradigms: bio-medical, psychological and socio-cultural. Most theorists (and practitioners), however, use a combination of them in analysing and describing the disorders.

According to bio-medical explanations, the causes of eating disorders range from malfunctioning glands to neurological factors and disfunctioning hypothalamus leading to overproduction or imbalance of hormones. Most recently, at least among bio-medical researchers, there has been a growing tendency to redefine causes formerly diagnosed as primarily social and psychological, to identify them as predominantly biological, namely genetic, factors. According to medical researchers at the University of Pittsburgh, for example, anorexia nervosa is a biologically-based illness linked to a malfunctioning of the brain chemical serotonin. However, they suggest as treatment a combination of medical prescriptions and psychotherapeutic interventions.[2] Moreover, bio-medical researchers and practitioners acknowledge that none of the purely bio-medical models has clarified the etiology of the disorder.

Psychological, including psychoanalytic, explanations vary as widely as the medical explanations. Most concentrate on identity and the girl's personal psychological development to maturity, focusing on the relationship between mother–daughter, father–daughter and other family members. Where they are not purely psychoanalytical in focus, most psychological explanations also incorporate reference to socio-cultural backgrounds, as it is perceived that societal influences have a role in the formation of the female subject (Schwartz, 1988).

In these explanatory models based in developmental psychology, eating disorders are viewed as identity crisis, as regression, as protest as well as submission, as signals thus belonging to the realm of blocked emotions, regressive behaviour and unresolved problems in the unconscious, all of which might result in syndromes or at least in a 'troubled relationship to food' (Minuchin, Rosman and Baker, 1978; Boskind-White and White, 1983; Chernin, 1985; Romeo, 1986; Gordon, 1990; Crowther, 1992).

The family – in particular the interrelationship of its members, its organisation as a 'private' sphere with specific codes of conduct and patterns of behaviour, including food patterns – is commonly perceived as one of the breeding grounds for eating disorders. From this perspective, the disorder then becomes an element of protest, or denial, or regression or terrorism by the girl towards the other members, mostly the mother and the father (Brumberg, 1988).

Within feminist therapeutic circles, the psychological analysis of the father–daughter interrelationship has received prominent attention, especially within the framework of child abuse and incest, but also within the framework of the daughter 'craving' for the attention in Western societies, of the mostly emotionally and physically absent father. According to the author of *Father Hunger*, if this 'normal craving' remains unacknowledged, it will lead to self-doubt, pain and anxiety and ultimately might result in conflicts with food and weight (Maine, 1991).

Socio-cultural explanations focus on aspects of socialisation of gender in Western societies, on constructions of sex and gender, on gender relationships and gender inequality (MacSween, 1993). In relation to eating disorders, the primary focus has been on the importance Western culture gives to the physical appearance of women and on the body shape currently seen as the female ideal: thin, tubular, fit and healthy. It is said that in Western culture, over several of the last centuries but especially since the 1950s, the ideal body image of women has become increasingly thinner. Women are surrounded by images of thin to super thin women, propagated by the media, films and fashion, health and fitness industries. Studies for New Zealand have shown that models in advertisements indeed are considerably thinner than they were in the 1950s and 1960s, correlating with a weight reduction of 9–10 kg. This means that models who are 175 cm (69") tall weigh 50–54 kg and models who are 170 cm (67") tall weigh 45–49 kg. According to the authors reporting these findings, these body shapes and weights have been presented in magazines for at least 15 years (Fay and Price, 1993: 31).

The primary objective of socio-cultural researchers dealing with changing body shapes is not to prove evidence of an increasing thinness of female models but to prove a link between the body shape portrayed in advertising and eating disorders. However, there is ambiguity surrounding the influence that images and ideals can have on people's behaviour and beliefs. Claims of a direct link between advertising images and eating disorders have so far remained unproven, along with statements such as 'advertising images are by nature persuasive' (Fay and Price, 1993).

Theorists following Foucault's thesis of the Western disciplined body, such as Susan Bordo, have argued that the relationship of eating disorders to ideology and bodily ideals is complicated, paradoxical and ambiguous. In her analysis of body, gender and power relations in Western societies, Bordo states that the body should not only be perceived as passive, disciplined, objectified and suffering, thus as the locus of control, but also as the locus of resistance. In terms of control, the importance of physical appearance – ideal body shapes – produces a relentless pressure to discipline, order and shape the body in specific, sexually connoted ways. Yet in terms of resistance, it remains possible that a woman who, for example, exercises her body, might through that process discover the self-confidence and strength to improve her work position or otherwise influence her situation positively (Bordo, 1993). Accordingly, anorexia nervosa might be perceived as a form of resistance: resistance against complying with the double and sexualised standards of modern Western society.

Whilst Bordo places eating disorders in a complex framework of intersecting cultural, historical, economic and familial developments and psychological constructions of gender, she does not question the 'existence' of eating disorders or girls' illnesses. On the contrary, in keeping with the authors mentioned above, she perceives the 'escalation' of eating disorders as a 'significant social phenomenon' (Bordo, 1993: 33).

A fourth explanatory concept can be found among socio-historians and medical historians. Despite the overwhelming concentration of psychological, psychoanalytical and medical analysis on the so-called 'modernity' of the illness, much of the most recent research has been influenced by a range of socio-historical studies (Bell, 1985; Brumberg, 1982 and 1988; Knibiehler, 1983; Loudon, 1980 and 1984; Shorter, 1987; Skrabanek 1983; Vandereycken and van Deth, 1994). These studies reveal two consistent findings: the act of hungering can be traced back to medieval, even classical times (holy anorexia) and girls and young women throughout history have apparently suffered from a range of 'maiden diseases'.

Maladies/afflictions referred to as maiden diseases – including morbus virgineus, love sickness, furor uterinus and green sickness – are defined as maladies or afflictions specific to young women and girls. According to socio-historical studies, maiden diseases have been part of philosophical, pedagogical, medical and cultural discourses since early modern times. The historical roots of these pathological constitutions, which are foremost sexual and gender specific, can be traced back even further. Classic Greek and Latin and medieval texts describe ailments which befell young women and were related to love, sexuality and the status of virginity. Explained from the Hippocratic and Galen humoral pathological framework, the ailments were referred to as disorders related to imbalance or malfunctioning of the fluids, blood, bile and (female) semen, to sexuality, longing and melancholy. Medieval and Renaissance doctors also used philosophical and cultural explanations, most probably because at the time the medical framework was built on a philosophy of wholeness in which body, soul, humours, temperament, environment, the earth and the cosmos were seen as interrelated (Liebault, 1632). Medicine had not yet become fragmented and specialised (Laqueur, 1990).

From the Renaissance and early modern times, parallel to its growth in professionalism and prestige, the medical profession began to give serious attention to maiden diseases. Further, from the eighteenth century a common term to describe maiden diseases became the more precise, medically defined 'chlorosis'; in English this translated as 'green sickness', in Dutch, French, German and Scandinavian as 'pale sickness' (Bullough and Campbell, 1980; de Ras, 1993a and b; Figlio, 1978; Horbach, 1989; Hudson, 1977; Loudon, 1980; Showalter, 1985; Siddal, 1982; Starobinski, 1981; Theriot, 1987).

From the eighteenth century onwards, doctors diagnosed chlorosis predominantly as a menstruation disorder in girls, such as ammenorhea or dismenorrhea. They also declared it to be most of all a disorder of the menarche, thereby relating it to female puberty.

In the first half of the nineteenth century, doctors noticed an alarming rise of chlorosis, an 'epidemic' which reached a peak in the middle of the century. However, with the discovery of anaemia as a form of iron deficiency, by the end of the century the diagnosis for chlorosis became an iron deficiency disorder. Though anaemia and iron deficiency were understood as non gender specific diseases, most physicians were convinced that chlorosis was a exclusively female form of anaemia in girls and young women. With the introduction of iron treatment, chlorosis more or less disappeared from the medical registers, although even today questions about its etiology arise.

Samuel van Hoogstraeten: The Anaemic Lady, Rijksmuseum, Amsterdam, c. 1670.

Just as chlorosis disappeared, another illness entered the medical registers which was likewise understood as a malady typical of young females: anorexia nervosa. In 1873 a French neurologist, Charles Lasègue, published an article entitled 'De l'anorexie hystérique' (Lasègue, 1873a and 1873b). In the same period a well known English doctor, William Gull, published 'Anorexia Nervosa' (Gull, 1874). Eating disorders were known under the name pica, or anorexia, which means hungering. Both doctors claimed that they had discovered the disease. Lasègue claimed he had been unaware of an 1868 address by Gull in which Gull had focused on 'hysteric apepsia', added the word 'nervous' to anorexia and declared it a form of hysteria, that other well-known grand female malady of the nineteenth century.

The etiology of anorexia nervosa proved to be as unclear as chlorosis had been, if not even more so. For, amidst the rising prominence of the Freudian unconscious, of psychology, psychoanalysis and psychiatry, the etiology of anorexia nervosa took shape as a 'nervous' disorder. However, although the disorder existed on the medical registers from the late nineteenth century, it was only in the 1970s that doctors started to perceive an increase in its incidence. Anorexia nervosa was to become the ultimate girls' disease of modern times.

In their analysis of the disorder, most socio-historical researchers emphasise the end of the nineteenth century as the era not only of the medical birth of anorexia nervosa but also of major emancipatory changes for women and girls. In her study of the historical specificity of anorexia nervosa, van 't Hof states explicitly that 1865 to 1880 was the period when girls throughout Western countries received, for the first time in history, official access to higher education. In her view, this socio-cultural event heightened social concern about girls. It was accompanied by extensive professional debate, writing and analysis, especially among educators, pedagogues and physicians. Those opposing the change stressed the detrimental effects of schooling and intellect on the 'health' and wellbeing of girls and young women, such as the loss of 'feminine' characteristics and, more importantly, the degeneration of their reproductive qualities. Well-known doctors explained the perceived rise of girls' diseases such as menstrual disorders and disorders with symptoms comparable to anorexia nervosa as a direct result of too much education, in which the brain and intellect were exercised (van 't Hof, 1994: 160). Van 't Hof underscores these parallel events to strengthen her thesis that it was not so much the disease anorexia nervosa which was born at the end of the nineteenth century but the concern and interest of physicians and other professionals regarding the wellbeing of girls and young women.

Brumberg (1988) focuses equally on socio-historical phenomena, namely girls and women gaining access to higher education and the rise of emancipation movements. She stresses, however, that traditional Victorian expectations, education and thus socialisation also continued at this time. There thus existed somewhat contradictory societal influences, with both the emergence of more space and liberty for young women and continued double moral standards of the bourgeoisie (which, it might be noted, were reinforced by treatises of medical and educational professionals). The extra pressures on young women that this situation created may have resulted in the emergence of anorexia nervosa.

Brumberg claims that eating disorders such as anorexia nervosa have appeared intermittently at different historical moments, within different societal and familial contexts and that each link should be closely analysed. The increase that occurred in the U.S.A. in the 1970s is, according to her, closely related to two aspects of modern American life: the relationship of society to food and to sex (Brumberg, 1988).

Psychologists, doctors and medical historians currently tend to use historical information to underscore their claim that anorexia nervosa is an existing and serious disorder. The medical historian Irvine Loudon has placed chlorosis and anorexia nervosa within the framework of the historical phenomena of girls' diseases. His focus has been the psychosomatic aspects of the putative ailments. He argues that chlorosis and anorexia nervosa are not two completely different illnesses which appeared one after the other, but two 'psychogenetic' manifestations of comparable structure which are directly linked to the passage from girlhood to adulthood. In this passage, according to Loudon, 'irrational' fears exist for the social and sexual implications of female adulthood, which might evoke attempts to avert the 'dramatic' pubertal bodily changes, such as the change in appearance and in weight. All this, according to Loudon, might result in 'regressive' behaviour such as the denial of food (Loudon, 1980 and 1984).

Some theorists and practitioners now assume that anorexia nervosa has *always existed* but that it appeared under different names and different diagnostic descriptions. Silverstein and Perlick, for example, claim that the so-called diseases of young women, such as chlorosis, hysteria and neurasthenia, are names given over the years to syndromes that have been said to afflict women at adolescence. According to the authors, these diseases were said to involve depression, anxiety, disordered eating, headaches and other symptoms. The authors do not believe that these disorders were identical: in their view, different disorders can result from differences between individual women such as genetic predispositions, family constellations, societal influences, or the age at which a woman begins to experience problems. However, they do believe that all disorders are most probably variants of one single disorder which they term 'anxious somatic depression' (Silverstein and Perlick, 1995: 33).

Furthermore, stating that gender bias plays a major role in causing psychological and physical problems, they conclude that historical research indicates that those women diagnosed under one of the categories of maiden disease were 'non-traditional females discontent with societal limitations placed on them' (Silverstein and Perlick, 1995: 161).

In short, socio-historical perspectives emphasise socio-historical and psycho-historical elements such as fear, stress and societal limitations as well as the gaining of societal spaces, traditional as well as non-traditional family settings, ambivalences girls faced in their traditional role as well as in new modern roles. All these factors are seen as causing the emergence of anorexia nervosa in the 1870s and its development into an epidemic in the 1970s. Apart from van 't Hof, none of these researchers questions whether this 'female' disease really did emerge, or indeed whether it reached 'epidemic' proportions.

Gotfried Schalcken, Doctor's Visit, 1660–70, Rijksdienst Beeldende Kunst, The Hague.

The History of Girls' Disorders as a History of Constructions. The 'Making' of a Disease

In the following paragraphs I move away from previous explanations to take a different angle on girls' illnesses and eating disorders. In particular, I focus on their construction, their 'architecture' and their being part of a wider and long-term socio-cultural and historical process.

In my view the discourse to date on girls' diseases can be read as waves of upcoming and disappearing preoccupations. At certain times, specific concepts and medical tags appear, such as maiden disease, love sickness and furor uterinus in the sixteenth and seventeenth century, chlorosis and/or green sickness in the eighteenth and nineteenth century, anorexia nervosa in the twentieth century and, most recently, bulimia nervosa.

As outlined above, in the sixteenth and seventeenth centuries girls' diseases begin to appear in medical, philosophical and literary texts, first as morbus virgineus or maiden disease. The phenomenon is explained, using a humoral pathological framework, as a disorder of the fluids or flow of blood and (female) semen. Sexuality is an intrinsic part of this framework. Then the phenomenon is named green sickness or chlorosis and its etiology becomes more specific to a menstrual disorder.

In the eighteenth century the phenomenon undergoes embourgeoisement: young women from the high bourgeoisie seem to be prone to all sorts of ailments. The term 'mademoiselle à vapeur' becomes a common expression. Meanwhile, the medical profession becomes more respectable, especially among the high bourgeosie. The ailment then becomes a classified, thus serious, disease.

In the nineteenth century the phenomenon undergoes various social and cultural changes. It enters other classes, especially the labour class in industrialised countries such as England. Chlorosis is perceived among many young female labourers who work long hours under extreme conditions and is declared a social illness. However, the disorder is also perceived among wealthy bourgeois girls, who, according to physicians, remain too long in their stuffy rooms (the 'girls' room' is an invention of the nineteenth century) and who do not exercise enough in the open air.

Until the second half of the nineteenth century an alarming rise in chlorotic girls is noticed and the ailment reaches 'epidemic' proportions. But, from mid-century, with the progress of the medical profession, the phenomenon undergoes rationalisation. Chlorosis becomes known as a female form of anaemia. Then the disease undergoes a phase of 'hysterisation' at the end of the nineteenth century as neurological and psychoanalytic explanations emerge. Sexuality, this time through the focus on psychoanalysis, re-enters the diagnosis. Biological and somatic explanations decrease, psychosomatic and social explanations increase. The trend continues parallel to the further development of the women's and youth movement and achievements of emancipation for women and girls.

In the twentieth century girls' diseases are further specified as eating disorders and as from the 1970s anorexia nervosa becomes the girls' disease par excellence. An increase and then an 'epidemic' are detected in the 1980s. Further specification occurs with ever more refined classifications of eating disorders. A new classification emerges: bulimia nervosa, the girls' disease of the 1990s.

In these historical 'waves' the following phenomena deserve our attention. First, with respect to eating disorders, starvation as a practice has a long history, not only among girls but also among adult women and men. However, it is only at the end of the nineteenth century that the medical profession has claimed it an illness, a 'nervous' condition, determined it 'female only' and ascribed it to puberty and adolescence. With respect to the architecture of an illness as typically female and as typically belonging to a certain life phase, there are more questions to be asked. Is anorexia nervosa indeed an illness which is typical for girls in puberty? Has anorexia nervosa indeed increased during the last 15 to 20 years as suggested by doctors, psychologists, pedagogues, feminists and the public?

Recent research found that an increase in anorexia nervosa could not be proved (van 't Hof, 1994). What could be proved, however, was a higher level of sensibility of doctors towards so-called girls' disorders. It appears that doctors increasingly diagnose all kinds of disorders in girls as a form of anorexia nervosa, an eating disorder. Furthermore, research has shown that starvation and fasting did and does not only appear in girls in puberty, but also in boys and in adults (Vandereycken, 1994; Wack, 1990; van 't Hof, 1994).

Second, with respect to girls' diseases in general, there is evidence which does not fit with the claim by socio-historians and medical historians that girls' diseases have always existed as afflictions typical of young women and girls and have peaked at certain periods in history. That is, it is notable that at least love sickness, which has become known as a typical girls' disease, was in the Middle Ages a prominent ailment of young men of the nobility. (Wack, 1990) As a 'male' disease it was seen as a noble and delicate malady with a certain cultural status, namely that of possessing refinement (being of nobility meant possessing a sensitive nature), cultural creativity (the art of writing love poems) and geniality. My own research shows that it was only from the sixteenth century that love sickness became predominantly ascribed to young unmarried women of the high bourgeois classes. It is important to note that from this time onwards, together with the declining status and decreasing numbers of the nobility, love sickness among young men became perceived as highly unfavourable and in medical terms was described as mental illness. In a parallel development, love sickness among young women and girls came to be perceived as an illness closely related to sexuality in a pejorative sense, to pre-marital pregnancy, which was morally and socially condemned and to menstrual disorders. As a female 'quality', love sickness had dropped considerably in its cultural and social ranking. In short, whilst male love sickness in the Middle Ages was perceived as evidence of belonging to a high social class and as culturally positive, love sickness among bourgeois girls and young women had overtly negative sexual connotations and was condemned as morally wrong, as socially and culturally problematic and as lowering the upcoming status of the bourgeoisie (de Ras, 1993b).

Third, most social scientists concentrate on the end of the nineteenth century as the point of the emergence of eating disorders and the end of the twentieth century as the point that it reached epidemic proportions. As discussed above, they link the 'discovery' of anorexia nervosa to major social, cultural and political shifts taking place which influenced the position of women and girls in society.

Yet is it so easy to connect the phenomenon of girls' diseases with social and cultural changes of, for example, the late nineteenth century? Since the emergence of medical treatises, medical concern with girls' ailments is coupled with concerns regarding girls' social, cultural, psychological and pedagogical background (de Ras, 1993a). The relation of girls' maladies to, for example, schooling and intellect can be found as early as the Renaissance. Moreover, socio-historians have found that from the seventeenth century a variety of educators, philosophers and the public generally have used medical arguments to counter the growing interest among girls and women in science, learning and intellect (Rang, 1985). Thomas Laqueur found that the former one-sex model in medical science changed into a two-sex model of man and woman in the eighteenth century – which was specifically the time that women started entering the public domain of politics, education, the arts and writing. According to him, this new medical and philosophical discourse of gender differences provided new arguments for confining women to specific female spaces, the private sphere in society, in order to counter their aspirations towards entering the public domain (Laqueur, 1990).

Furthermore, it is useful to take into account research which concentrates on socio-cultural habits in history relating to health, such as food and diets but also medical practices. In the case of chlorosis, social theorists who accept that it was caused by iron deficiency, have claimed that iron deficiency in girls and young women was most probably reinforced by the medical treatment of blood letting, a common and increasingly popular treatment until the mid-nineteenth century. They see their argument as further strenghtened, given that chlorosis was diagnosed as a menstruation disorder, by the broader social context of malnutrition and poor diets. From an analytical perspective on the 'making' of a disease, such findings on socio-cultural habits at least problematises bio-medical and socio-cultural explanations, in this case of chlorosis.

In my view the discourse of 'scientia sexualis' has, from the Renaissance and Humanism onwards, constructed a causal link between the 'nature' of woman and characteristics of femininity, between female sex and female gender. This discourse – part of early modern and modern Western philosophy, the natural sciences, medicine, pedagogy and psychology – has inextricably linked 'female maladies' to biological, sexual and bio-psychological constitutions defined as 'female'. From the nineteenth century, the link was emphasised following changes to the medical profession: its rationalisation and professionalisation, its rising prestige and its concentration on discovering, classifying and curing diseases. Against this background and to enable the medical profession to prove the link between female maladies and constitution, female illnesses had to be perceived, classified and made visible as specifically female.

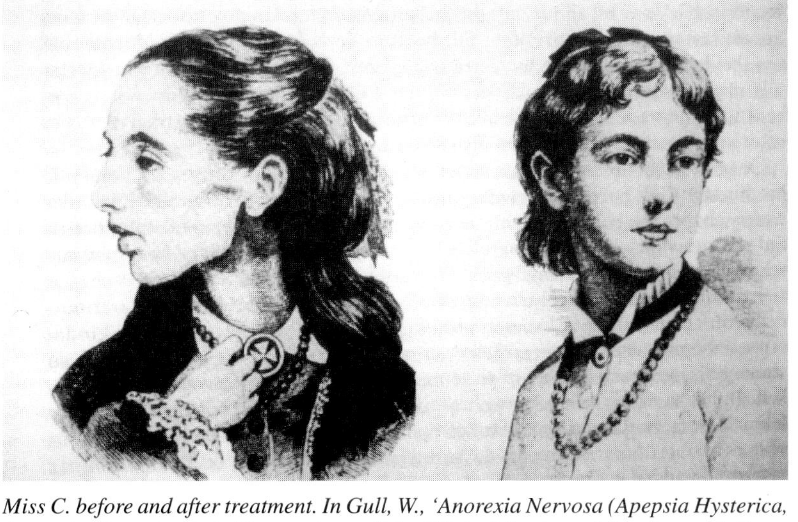

Miss C. before and after treatment. In Gull, W., 'Anorexia Nervosa (Apepsia Hysterica, Anorexia Hysterica)', in Transactions of the Clinical Society of London, *vol. 7, pp. 22–28.*

In this respect the invention of photography at the end of the nineteenth century – appearing amidst and probably because of, an era when Western societies became increasingly scopophiliac, thus oriented towards the visual and representative – formed a new and important 'tool' for physicians. As O'Connor has elaborated in his analysis of the relationship between the visualisation of anorexia nervosa and its emergence, William Gull accompanied the announcement of his discovery with a set of photographs (O'Connor, 1995). These photographs showed anorexic patients before and after (successful) treatment. O'Connor suggests that because it was hard to find any somatic cause for the illness, the photographs 'proved' that a distinct medical disorder existed. The visual representation of anorexic bodies became a centrepiece for doctors who diagnosed their patients as suffering from this new disease and subsequently proved their diagnosis by using photographs: see, here is the emaciated body which I diagnosed as anorexia nervosa. More than that, photographs became the tool to show the success of the medical profession: see, here is how she looked before treatment and here is how she looked after my successful treatment. The photographs allowed the physicians to dramatise the illness through the spectacle of emaciated girls' bodies. It was a spectacle not unlike the spectacle of that other grand female malady: hysteria.

Conclusions

From the viewpoint that the construction of girls' illnesses is part of a wider, long-term socio-historical process, I perceive in the medical, psychological, educational, social and political (also feminist) discourse a growing *specification* of girlhood as a psychosomatic life phase. Female puberty and adolescence in particular, are perceived as a precarious life phase which is easily outbalanced, is unstable *per se* and is under the constant influence of societal, biological, sexual and psychological pressure.

It is, in my view, not the illness itself which took on epidemic proportions in the late twentieth century, but the discourse of specification, the concentration by doctors, educators, psychologists, feminists, parents and the public on girls' puberty and adolescence. By specifying this life phase, they have constructed it as different from boys' puberty and adolescence and different from adulthood. The nexus of this discourse is an overall concern for the bodily, sexual, mental and especially moral, 'wellbeing' of girls. Medicalisation is but one cornerstone in this process of specification of female puberty and adolescence. As I have argued elsewhere, the process is specific to time, society and culture and is thus related to specific socioeconomic, political and cultural transformations and events in Western history (de Ras, 1993a).

This historical specificity can be traced to the emergence of pubertal and adolescent girlhood in the fifteenth and sixteenth centuries. It was then, for the first time in Western countries, that girls came to be perceived as a semi-autonomous group of female youngsters. This 'perceiving' was related to the appearance of young marriageable girls in public, schools and the labour market in the urbanised regions of Renaissance Europe, particularly the north-western regions, which was a direct consequence of economic, political, philosophical and religious transformations.

Their appearance as a group of female youngsters can be found in social and economic records such as statistics of age, school attendance, marriage contracts, guilds and occupations. Further, as a category and representation, they appeared in various cultural expressions: in language, texts, paintings and theatre. The word 'girl', for example, emerged for the first time in the sixteenth century. Finally, by becoming 'visible' in discourse and everyday life, the life phase of girls in puberty and adolescence was also 'created' and took shape as a social reality.

There is, then, considerable evidence in support of a theory of the making of a disorder; one specific to girls and young women in a particular historical, socio-cultural context. Yet theorists – be they physicians, medical historians, psychologists or psychoanalysts, psychiatrists, socio-culturalists or socio-historians, feminists or feminist therapists – have not asked if eating disorders are a medical construction, if they could be part of a process which I would call the pathologising of female puberty and adolescence. Instead, where they have used an analytical perspective of medicalisation, theorists have used it to argue that the illness is not biological or somatic but instead a direct result of the Western, patriarchal society, of socialisation in a world where gender inequality still exists. Hypotheses vary from statements that girls are oppressed, victims of abuse and of a body shape ideology, have too little space or too much space; to statements that Western society is built on contradictions between disciplining the body and freeing it, thus leaving girls and young women with an increasing multitude of dilemmas, the result of which is that they increasingly stop eating, or binge and purge. Amidst this broad range of claims, no theorist has questioned the aspect of 'vulnerability' of female puberty and adolescence. Indeed with every new text, theorists, practitioners and the public confirm that female puberty and adolescence is, by definition, a precarious, vulnerable phase, prone to imbalance, fear, anxiety and regressive behaviour.

I therefore would like to finish with some provoking questions:

Are girls in puberty and adolescence, biologically, psychologically and sexually so imbalanced, unstable and fragile (compared to boys) that they cope less with the pressure of society, the family and their developing bodies and therefore are particularly prone to eating disorders? Or: Do physicians, counsellors, psychologists, pedagogues and parents have a problem with girls' puberty and thus classify every behaviour, every problem considered odd, irregular or unfeminine as a disorder? And does the 'epidemic' of representation of this disorder, in research, analysis, texts, therapy, photographs, films, magazines and television contribute to the growing unease, not of girls and young women towards society, but of society towards girls and young women?

Notes

1. Bulimia nervosa appeared in 1980 for the first time in the *Diagnostic and Statistical Manual*, published by the American Psychiatric Association (DSM III). Since then, health professionals have shown a growing concern with this new disease.
2. December 1995 press release on the Internet of a study by researchers at the University of Pittsburgh Medical Centre's Western Psychiatric Institute and Clinic, also reported in the November 1995 issue of the *American Journal of Psychiatry*. http: //www.wpic.pitt.edu/press/press.htm.

References

Bell, R.M. (1985), *Holy Anorexia*, Chicago: The University of Chicago Press

Bordo, S. (1993), *Unbearable Weight. Feminism, Western Culture and the Body*, Berkeley, Los Angeles, London: University of California Press.

Boskind-White, M. and White, W.C. (1983), *Bulimarexia. The Binge/Purge Cycle. W.W. Norton and Company,* New York, London.

Brumberg, J. (1982), 'Chlorotic Girls, 1870–1920: A Historical Perspective on Female Adolescence', in *Child Development*, vol. 53, pp. 1468–1477.

Brumberg, Jacobs J. (1988), *Fasting Girls. The History of Anorexia Nervosa*, New York: Plume.

Bullough, V. and Campbell, C. (1980), 'Female Longevity and Diet in the Middle Ages', in *Speculum*, vol. 55, no. 2, pp. 317–324.

Chernin, K. (1985), *The Hungry Self. Women, Eating and Identity*, New York: Harper and Row, Publishers.

Crowther, J.H. *et al.* (1992), *The Etiology of Bulimia Nervosa: The Individual and Familial Context*, Washington, Philadelphia, London: Hemisphere Publishing Corporation.

Fay, M. and Price, C. (1993), *Eating Disorders and the Weight and Shape of Media Images of Women. Female Body Shape in Print Advertising.1953–1988*, Marketing Discussion Papers No. 7, University of Otago.

Figlio, K. (1978), 'Chlorosis and Chronic Disease in Nineteenth-century Britain: The Social Constitution of Somatic Illness in Capitalist Society', in *Social History*, vol. 3, no. 1, pp. 167–197.

Foucault, M. (1978), *The History of Sexuality. Volume 1: An Introduction*, New York: Pantheon Books.

Gordon, R.A. (1990), *Anorexia and Bulimia. Anatomy of a Social Epidemic*, Cambridge Massachusetts: Basil Blackwell.

Gull, W. (1874), 'Anorexia Nervosa (Apepsia Hysterica, Anorexia Hysterica)', in *Transaction of the Clinical Society of London*, vol. 7, pp. 22–28.

Hof, S. van 't (1994), *Anorexia Nervosa: The Historical and Cultural Specificity. Fallacious Theories and Tenacious 'Facts'*, Lisse: Swets en Zeitlinger.

Horbach, H. (1989), 'Die Behandlung der Amenorrhoe in der zweiten Hälfte des 19. Jahrhunderts, Insbesondere bei Mädchen im Pubertäts- und Jugendalter', in *Würzburger medizinhistorische Mitteilungen*, vol. 7, pp. 299–312.

Hudson, R.P. (1977), 'The Biography of Disease: Lessons from Chlorosis', in *Bulletin of the History of Medicine*, vol. 51, pp. 448–463.

Knibiehler, Y. (1983), *La femme et les medecins,* Paris: Hachette.

Laqueur, T. (1990), *Making Sex, Body and Gender from the Greeks to Freud*, Cambridge: Harvard University Press.

Lasègue, E.C. (1873a), 'De l'anorexie Hystérique', in *Archives Générales de Médecine*, vol. 21, April, pp. 385–403.

Lasègue, E.C. (1873b), 'On Hysterical Anorexia', in *Medical Times and Gazette*, vol. 6, pp. 265–266 and 367–369.

Liebault, J. (1632), *Trois livres appartenant aux infirmitez et malades des femmes*, Paris.

Loudon, I. (1980), 'Chlorosis, Anaemia and Anorexia Nervosa', in *British Medical Journal*, vol. 281, pp. 1669–1675.

Loudon, I. (1984), 'The Diseases Called Chlorosis', in *Psychological Medicine*, vol. 14, pp. 27–36.

MacSween, M. (1993), *Anorexic Bodies. A Feminist and Sociological Perspective on Anorexia Nervosa*, London and New York: Routledge.

Maine, M. (1991), *Father Hunger. Fathers, Daughters and Food*, Carlsbad, CA: Gürze Books

Martin, E. (1994), *Flexible Bodies. Tracking Immunity in American Culture From the Days of Polio to the Age of AIDS*, Boston: Beacon Press.

Minuchin, S. Rosman, B.L. and Baker, L. (1978), *Psychosomatic Families. Anorexia in Context,* Cambridge, Massachusetts and London, England: Harvard University Press.

O'Connor, E. (1995), 'Pictures of Health: Medical Photography and the Emergence of Anorexia Nervosa', in *Journal of the History of Sexuality*, vol. 5, no. 4, pp. 535–572.

Rang, B. (1985), 'Am Rande des Wahnsinns. Pädagogische und Medizinische Kritik Weiblicher Gelehrsamkeit im Zeitalter der Aufklärung', in *Traditio Campadis. Das Versöhnende dem Zerstörenden Abtrotzen*, Liechtenstein: Topos Verlag.

Ras, M.E.P. de (1993a), 'Vrouwelijke Jeugd. Gender en Levensfase als Thema van Historisch en Cultuursociologisch Onderzoek', in *Van Frictie to Wetenschap. Jaarboek 1992–1993 Vereniging van Akademie–onderzoekers*, vol. 4, pp. 65–74.

Ras, M.E.P. de (1993b), 'An Iconography of Dutch Girlhood', in *Girls, Girlhood and Girls' Studies in Transition*, Ras, M.E.P. de and Lunenberg M. (eds), Amsterdam: Het Spinhuis Publishers, pp. 337–356.

Romeo, F.C. (1986), *Understanding Anorexia Nervosa*, Springfield: Charles C. Thomas Publishers.

Schwartz, H.J. (ed.) (1988), *Bulimia: Psychoanalytic treatment and History*, Madison Connecticut: International Universities Press.

Shorter, E. (1987), 'The First Great Increase in Anorexia Nervosa', in *Journal of Social History*, Fall, pp. 69–96.

Showalter, E. (1985), *The female malady. Women, madness and English culture, 1830–1980*, New York: Penguin.

Siddal A. Clair (1982), 'Chlorosis – Etiology Reconsidered', in *Bulletin of the History of Medicine*, vol. 56, no. 2, pp. 254–260.

Silverstein, B. and Perlick D. (1995), *The Cost of Competence. Why Inequality Causes Depression, Eating Disorders and Illness in Women,* New York: Oxford University Press.

Skrabanek, P. (1983), 'Notes Towards the History of Anorexia Nervosa', in *Janus. Revue internationale de l'histoire des sciences, de la médecine, de la pharmacie et de la technique*, vol. 70, pp. 109–128.

Starobinski, J. (1981), 'Chlorosis – The "Green Sickness" ', in *Psychological Medicine*, vol. 11, pp. 459–468.

Theriot, N.M. (1987), 'Psychosomatic Illness in History: The "Green Sickness", Among Nineteenth-century Adolescent Girls', in *The Journal of Psychohistory*, vol. 15, no. 1, pp. 461–480.

Vandereycken, W. and Deth, R. van (1994), *From Fasting Saints to Anorexic Girls. The History of Self-Starvation*, London, Atlantic Highlands NJ: The Athlone Press.

Wack, M.F. (1990), *Lovesickness in the Middle Ages. The Viaticum and its Commentaries*, Philadelphia: University of Pennsylvania Press.

Part 3

Reading the Body as Text

Autopsy as Gaze

The Construction of the Hermaphroditic Corpse as a Text of Sexual Difference

Susan Duncan

Picture a surgical ward that specialises in plastic surgery. This is the ward where the babies come who aren't really obviously boys or girls. These babies are described as ambiguously gendered or even as hermaphroditic. The families of these babies are under immense stress, they do not have the answer to that very first question they are inevitably asked: 'What did you have?' There are after all only two choices with which to answer this question. One either has a boy or one has a girl. The difference can of course be read from the baby's genitals. Remember all those movies you have watched where the attending doctor or nurse and in these more family centred times the father, looks at the baby's genitals and announces to the eagerly awaiting and exhausted mother, 'It's a girl/boy!'

There is a difference and the genitals are where it can be read. The babies that come to this ward though have genitals that can't be read that easily.

The babies are here to be surgically assigned their correct gender, the ambiguity of their genitals is seen somehow to be a mistake. There is something wrong here that is seen to necessitate surgical intervention. The babies are admitted, undergo surgery and are discharged whereupon the proud parents can answer the question of what they have had. The child is deemed to be either male or female so they can take their place in the world. What place has the infant that is neither male or female, I asked myself.

In a text book that tells the story of surgical gender assignment and reassignment, I read of the criteria that determine whether one is a boy or a girl. It contains the story of identical twin boys one of whom was to go to a ward like the one I mentioned above in America in the seventies. John Money (1973: 293–294), the author of the chapter from which I read is at the time the Professor of Medical Psychology and Associate Professor of Pediatrics at the John Hopkins University School of Medicine. The text is accompanied by photographs of the twins. This is what it says:

> In Figure 59 you have twins whose embryogenesis and fetal development was normally male. The sex reassignment of the

> one was made because of an accident. You see (Figure 60) that it was truly a total accident. There is no penis left at all. It was lost as the result of a miscalculation in the use of an electric cautery instead of a blade for circumcision. A burn killed the tissue off the penis, which, unable to heal, withered and dropped off.
>
> The child was sex-reassigned after tremendous agonizing on the part of the parents.

The presence or absence of a penis, in this case seems to determine whether one is deemed to be a boy or a girl. I asked myself why is it that the decision made for that little boy changed him into a little girl. This question took me many places, into the fascinating realms of theory(ies). I want to elaborate on this notion that sexual difference is read from the body. In this respect the body becomes a text from which sexual difference can be read. There are boy genitals and girl genitals and the difference between these two is thought of as distinct and eternal. Infants born whose genitals are unreadable as one or the other problematise this eternal difference between male and female. The either/or nature of male and female here becomes blurred. I believe that this is the reason that the 'correct' sex is sought and assigned within Western medical practice. This search for the truth of sex does not permeate all places and all times. This necessity for one to have a true sex is a relatively recent historical phenomenon. It is also implicated in the curiosity that surrounded hermaphroditic corpses within nineteenth century Western medicine.

> And when dire necessity at last required positive aid to relieve his sufferings, astounding as it may appear, he nevertheless essayed in struggling efforts to battle with those who kindly tendered him assistance, cursing in bitterness and blaspheming with strange inarticulate oaths even until the exacerbations of disease had overwhelmed his functions. Such was his horrible dread of being handled or rendered a spectacle subservient to the gaze of others.
>
> (James Akin, 1839: 4)

The above citation from a pamphlet published in 1839 describes the 'Facts' connected with the life and death of James Carey, a man reputed to be an hermaphrodite. Carey died in 1839 knowing that his corpse would subsequently be subjected to the medical gaze. This gaze read bodies of those deemed hermaphroditic as texts that could, in their difference from normative bodies, define and anchor a dualistically gendered sexual difference. In this way, the corpses of people whose bodies were inscribed as different to the originary norm serve as the primary text from which to read this very difference. Difference thus becomes a retroactive construct.

Within this chapter I describe the hermaphroditic corpse as being inscribed and *produced* as a text of sexual difference.[1] By this I mean that the distinct and eternal nature of the difference between male and female is held in place by a chimeric other. In order to establish an identity there must be a boundary established between the two mutually constructive terms. We can liken this to the continuous production and reproduction of self and other when what is Other to the self has to be recognised as such and delineated off from the self. In this respect it is the self who partially effects the delineation of the boundary. The other also holds power in the production of identity in that identity production between self and other is not simply a one-way cause and effect relationship. Rather, the relationship is dialectical. There is, however, a power differential operating wherein the self as subject is in a position of greater power to define the boundaries of inside and outside that determine the identity of both subject and object. The other then is a partial construction of the self.[2]

The medical gaze fixes and classifies the hermaphroditic corpse to effect classification of dualistic sexual difference. I am here deploying Foucault's notion of the gaze as described in *The Birth of the Clinic* and *Discipline and Punish*. The writing of Jacques Lacan is also relevant to the conceptualisation of a gaze which results in the production of a differentiated object. Within *The Four Fundamental Concepts of Psycho-analysis*, Lacan describes this concept of the gaze. For Lacan the gaze is the underside of consciousness. The gaze, in this respect, is constitutive of the subject as it identifies the object that is outside it and suspends itself from this different delineated outside. Thus the subject *is* via the gaze. By this I mean that the subject comes into being via the differentiation effected by the gaze; in order to constitute itself, the subject must separate itself off from that which it is not. This is done in relation to an object that is different and therefore separable.

In relation to the medicalisation of sexual difference I am arguing that there is always a gaze behind anatomical inscription and description. In this respect, the male medical gaze is one that seeks to define the heterosexualised duality of male/female as subject in relation to that which is different and separable as an object, the hermaphroditic body.

To illustrate this *production*, I will be paying particular attention to the life and spectacular exhibition of Gottleib Gottlich, a nineteenth century hermaphrodite '*produced*' by the medical gaze. Following the discussion of Gottleib Gottlich, I will move on to specifically concentrate on the production of the corpse as an anatomical object. Through the examination of corpses, the practice of autopsy provides access to the hidden depths where anatomical secrets dwell. Once I have established that the corpse is produced as an anatomical object of study within a discourse that effects constructions of masculinity and femininity, I will move on to argue that it was to corpses of difference that the medical gaze applied its mechanisms of classification and exploration. In order to do this I will look specifically at autopsies performed in the nineteenth century upon people whose bodies were deemed to hold hidden in life the secrets of sexual difference.

The sexually enigmatic status of the bodies of those marked as hermaphroditic was a source of confusion and curiosity. As it became more and more important to belong to one sex or another, sexes were clearly and medically delineated. The ambiguity written on hermaphroditic bodies confounded the boundary that was medically inscribed as the difference between male and female. Even under a thorough examination whilst still alive, the hermaphroditic body held its secrets in its depths – the depths of the body. The only way these secrets could be revealed to the medical gaze was on the autopsy table. There then developed a somewhat prurient desire to delve into the depths of the hermaphroditic corpse. It was the ambiguity of the bodies of those deemed hermaphroditic that was problematic and it was from the depths of the corpse that the 'truth' of this ambiguity could be read.

Post-mortem examination was the ideal situation where the body could be read as text. As is evinced by the opening citation regarding James Carey, the inevitability of this process caused great consternation. Despite the wishes of those whose bodies became text, they were opened, read and often displayed. Corpses of difference, in effect, became the property of medical researchers. They also became a foundation for public knowledge and public display.

In my concluding argument I will argue that hermaphroditic bodies are produced as they are simultaneously denied their existence. In this respect, they are the necessary Other to an 'originary' embodied sexual difference. Hermaphroditic corpses, then, function as chimerical Other in order to anchor the dualistic heterosexualised couplet of male and female. This suggests that the originary nature of sexual difference is itself produced.

The Inscription of the Hermaphroditic Corpse as a Text of Difference

> The scalpel should be held somewhat after the manner of a writing pen.
>
> (Andrew Fyfe, 1826)

Within Western medical discourse the human body has not always been read as a text that describes sexual difference. The Western text(s) from which sexual difference was read was historically both philosophical and biblical. There has been a transition, then, from text as text to body as text. Let us take some time now to look at this change in location of what serves as the primary text. In doing this, we are of course problematising the notion of primary text itself.

In the concluding chapter of *The Western Medical Tradition 800BC to AD 1800,* Michael Neve (1995: 78) writes, 'To enter the historical arena is to enter a world where we see what we assume has always been present actually being manufactured, being created in political circumstances, in educational contexts, even in the marketplace, where we may think doctors in the modern sense, are out of place'. I want to now enter the historical arena of anatomy to illustrate this assertion of Neve's by exploring how that which appears universal and natural in our world has been discursively produced.

Within anatomy it is a commonplace understanding that there is a direct link between the human body and its subsequent representation and description. Indeed, Bernard Knight (1980: 166) opens his text *Discovering the Human Body: Corpus Delicti – A History of Human Dissection,* with the following statement, 'Anatomy is a totally descriptive subject, thus its quality can only be as good as its basic material'. Anatomy is, in this respect, a purely descriptive subject that is inseparable from and totally dependent upon the human body as the originary location of subsequent knowledge. It is this assumption that I want to problematise by examining the history of the discipline. I will argue that the primary text of anatomy was initially philosophical and theological. There is, in this respect, a degree of independence between the body and the knowledge generated concerning the body when the philosophical text was the primary text for anatomical knowledge. Anatomical dissection was performed in order to support the philosophical text. The body in its materiality is here functioning as a secondary source. Following this discussion I will move on to examine the transition wherein the body becomes the primary text for anatomy. I will use the work of Andreas Vesalius, widely recognised as the father of modern anatomy, to illustrate the challenge to the text as the founding document of medicine. I will however, argue that anatomy is not simply a product of the body but also simultaneously produces it. This process of inscription serves to ensure that the body is always already coded with culturally specific meanings of sex.

In *A Short History of Anatomy from the Greeks to Harvey*, Charles Singer (1957: 7) establishes Aristotelian philosophy as the basis of all subsequent understandings of biological development. The influence of Aristotle extends for 2,000 years, including his philosophical text regarding 'the body' and embodied sexual difference. In effect, Aristotle named and codified the body through his philosophical text. Indeed, as pointed out by Singer (1957: 18), the nomenclature of the uterus used today has been retained from Aristotle's philosophy.

It is an accepted certainty that Aristotle's philosophical text has had an immense influence on the subsequent development of Western medicine. It is also a 'tolerable' certainty that Aristotle never dissected the human body. The philosophical writings of Socrates and Plato concerning anatomy were not based on dissection either. Singer (1957: 23) goes on to describe Aristotelian philosophy as 'the main intellectual pabulum of mankind'. Knowledge produced from this intellectual 'nourishment' was a knowledge that was not based on the material reality of 'the human body'. Rather the philosophical text provided the ontological basis of knowledge concerning, among other things, sexual difference. There is an independence here between anatomical structures and the knowledge produced concerning them.

Galen of Pergamum (129 to c. 216) is identified by Vivian Nutton (1995b: 58) as arguably the central figure in the development of the Western tradition of medicine. Galen believed that a good doctor should also be a good philosopher. This link between medicine and philosophy is a very strong one that extends well into the seventeenth century. In the Galenic system anatomical findings served to confirm rather than produce the truths of philosophy. The anatomy of the body then serves as a supportive text to the primary text of philosophy.

Galen's medical writings were written within an Aristotelian philosophy. Together they combined to form a medical textual canon that was to hold sway over anatomical knowledge for centuries to come. The study of medicine developed wherein the authority of Galen and Aristotle was unquestioned. Galen's anatomical writing was based not on the human body, but upon a reliance on animals for anatomical dissection. The explanation for this is attributed to what Nutton (1995d: 176) describes as 'the Aristotelian ladder of nature, with rational man at its summit and with the consequences of this doctrine, that in its material aspects the human body could be described and explained like that of a pig'.

There emerged then a textual canon from which knowledge concerning anatomy and medicine was gained. With the rise of the universities this literary textual tradition was strengthened. Medical education concentrated on written texts to the extent that Lawrence Conrad (1995: 131) asserts that, 'many students must have entered medical practice without even having treated a patient'. The location of medical faculties within universities served to ensure that written text remained the primary text for anatomy.

The written text of Aristotle and Galen combined to found an anatomical tradition that was grounded in written text. The proof for claims made by medical practitioners was located in the texts of Galen and Aristotle. Medical authority was textual. How the body functioned could be determined via a close examination of the writings of Galen. The material reality of the body was not the source for such information. Authority for claims was conveyed by text not embodied materiality. Dissection was, however, practised in antiquity. Corporeality and its display via dissection served to confirm the unquestioned authority of the Galenic canon. Indeed, Nutton (1995d: 179) speaks of, '... the linkage of the transient body to the more permanent words of a text ... [where] The visual evidence of the body supplemented the doctrines of the text'.

The ontological ground for anatomical knowledge then was words. Nutton further illustrates this with his discussion of the transmission of medical knowledge through encyclopaedias like the *Etymologies* of Isadore of Seville. Nutton (1995c: 85) states, 'Its very title, though, suggests the new context in which its medicine was studied – as words, by grammarians. To understand the meaning of the terms was to penetrate to the very essence of the divinely created world'. Contradictory evidence need not necessarily indicate an error within the Galenic system. Rather, as Nutton (1995d: 146) points out, 'contradiction could be reconciled by logic, by words and by a better understanding of the primary Galenic text'. A better understanding of the body could not therefore reconcile contradiction between body and text.

Dissection served, then, as a means of verification of an unquestioned authority. Dissection was a spectacle wherein the authority of text was reinforced. The human body is in this respect not unlike a supportive text. There is still a primacy given to the word. The body did not serve to reveal its inner mysteries but was rather placed secondarily within the wider philosophical canon of medicine. In this respect what could be seen, visually supplemented the text. How, then, did the human body become what Thomas Laqueur (1990: 70) describes as 'proof text' for claims regarding sexual difference? Ludmilla Jordanova (1989: 51) also deploys this conception of the body as text. She describes the human body as having a legibility that could be read. There was a transition where the location of anatomical truth changed from that of written text to the opened autopsied human body as text.

Transition of Primary Text

Andreas Vesalius is widely recognised as the founding father of a modern anatomy based on the ontological sureness of the human body. It is through the work of Vesalius that the body is to replace literary text as the primary text of anatomical knowledge. Indeed Sander Gilman (1989: 71) in his *Sexuality an Illustrated History* states, regarding Vesalius on Galen, 'Galen, writes Vesalius, never dissected a human uterus. He, Vesalius, will expose the errors of mere book learning through the act of opening the female book'. Through dissection and autopsy Vesalius will open the female book/body and from there read her truth.

For Vesalius, then, observation of the human body was the means by which one could gain knowledge concerning anatomy. To get information and knowledge concerning the structure of the human body, one then had to see for oneself rather than rely solely on the word of authority. Andrew Wear (in Conrad *et al.* 1995: 291) in his discussion of Vesalius makes the observation that the word autopsy comes from *autopsia,* which means seeing for oneself. Vesalius consistently and systematically tested the word of Galen against the primary text of the body. In his text *de fabrica,* Vesalius corrected over 200 'errors' found in the Galenic text (Knight, 1980: 177). The primary text of anatomy had become the ontological certainty of the text of the body.

Vesalius opened the female book and from there read her truth. However, as noted by Singer (1926: 108), there are structures present in Vesalius's text that are found in the Galenic tradition and not the human body. Vesalius read from the text of the body. The text from which he read had in some respects been written via Galenic text. This is particularly apparent in respect to Vesalius's representation of the female system described by Singer (1957: 130) as medieval and full of errors.

Vesalius was not alone in seeing in the body things that were present in Galenic text. According to Singer (1957: 102), the sixteenth century anatomist Charles Estienne 'claims to have seen with his own eye all that he describes. For this, however, he must have had the eye of faith, for he describes structures found in the text of Galen, not in the human body'.

The human body was inscribed as text through the text of Galen. It is therefore not surprising that anatomists, when reading this inscription, saw what they saw. What is remarkable is that they read it as the primary text. The body in its originary state of nature has replaced books and their authority. The culmination of the transition comes with the work of the seventeenth century anatomist Charles Harvey who states, 'I profess to learn and teach anatomy not from books but from dissections, not from the tenets of Philosophers but from the fabric of nature' (as cited by Wear in Conrad *et al.* 1995: 337).

Autopsia has replaced the word as the authority on which anatomy rests. The anatomy of the human body does not, however, reside in an originary state of nature. What is read from the book is written into the book simultaneously. By this I mean that the corpse is inscribed as text. It is this inscription which is subsequently read. Therefore even though the human body, made accessible through death and autopsy, forms the primary text of anatomy it is a text that remains enmeshed in the matrix within which the fabric of the body is woven. This locatedness of the body denies the possibility of anatomy ever being purely descriptive.

Autopsy, seeing and naming for oneself, carries with it simultaneously inscription and description. In this respect Vesalius, as Galen before him, has left his name inscribed within the body. The body as it is read in its ontological concreteness has been inscribed and described. The body has been produced.

Michel Foucault (1980: 115) closes Herculine Barbin's memoirs by stating, 'In the month of February 1868, the corpse of Abel Barbin, who had committed suicide by means of a charcoal stove, was found in a room in the quarter of the Theatre de l'Odeon'. A full autopsy was performed on the corpse of Herculine Barbin. As cited by Foucault (1980: 129), the physician writes, 'I was informed about this situation by Doctor Duplome, who regretted as well as myself that science was going to lose the opportunity to make a study of it and together we requested Doctor Regneir to use all his influence with the police superintendent so that I might be permitted to perform the autopsy and to remove the different parts on which the anomaly had a bearing'.

Doctor Reigner's comment illustrates the prurient scientific curiosity for corpses of difference. It is the anomaly present in the corpse of Barbin that makes it so attractive to science. By removing all the 'parts' marked as not normal, Reigner can further seek to define the norm. This is achieved by defining the norm in its necessary difference to the anomalous. In this respect, the norm and the anomalous can be described as *mutually productive*. Therefore the primary originary norm has already defined the difference before it is read from the corpse of Barbin. This suggests that the primary term constructs a secondary term to reflect itself in its difference. In this respect Barbin's difference is written before it can be read. This is of course quite normal as in order to read something it must, of necessity, first be written. Barbin's corpse is therefore unreadable outside of the matrix that has produced it as different. But it is precisely in this difference that it is read.

Herculine Barbin was a nineteenth century French hermaphrodite whose corpse was read as a text that detailed his or her difference from the sexually dualistic norm. In this respect, Barbin's 'deviant' status as hermaphrodite provides knowledge against which the 'norm' of a dualistic sexual difference can be constructed and known. Barbin's corpse is then a text from which its difference from heterosexualised normality can be read. His or her corpse is a text of difference.

The Medical Gaze as Productive of Identity Through Difference

I want to now enlarge on the argument that the medical gaze effects identity through the construction of difference in regard to masculinity and femininity. First, I will argue that the medical gaze constructs, as its object of study, a dualistically differentiated and sexualised couplet of bodies. With this couplet are the mutually exclusive categories of masculine and feminine. Second, I will argue that hermaphroditism functions as a chimerical Other to this couplet. I shall use stories told concerning people whose bodies were deemed to be ambiguously sexed to illustrate these themes.

Instead of being ahistorical certainties, masculinity and femininity are the results of historical, social and cultural processes of politicised construction, reconstruction and, of course, deconstruction. It is becoming a critical commonplace to assert that this construction occurs at the level of bodily materiality.[3] Denise Riley (1988: 101) identifies the female body as the 'obstinate core of identification of women'. She goes on to state, ' "The body" is not, for all its corporeality, an originating point nor yet a terminus; it is a result or an effect'. For Riley, 'the body' necessitates marking with time and sex. She takes this step and asserts that this is still not enough. Riley (1988: 103) continues,

> If female bodies are thought of as perennially such, as constant and even embodiments of sexed being, that is a misconception which carries risks. ... It could be claimed that a characteristic of

> the sadist's gaze is to fix and freeze its object, to insist on absolute difference, to forbid movement.

Riley, in saying that the perceived constancy of embodiment is a misconception, portrays the gaze through which absolute difference is constructed as a sadistic gaze. In this respect, the insistence on absolute difference could be described as corporeal violence. This violence denies the mutability of corporeality. There is a resonance here with Michel Foucault's discussion of the medical gaze in *The Birth of the Clinic*.

For Foucault, the medical gaze is implicated in the construction of the objects it subsequently seeks to know. It is in this respect that the human body is constructed as an object by the medical gaze. There is *mutual* production involved here between the subject and object of study. Foucault (1973: x) states, '... thus the articulation of medical language and its object will appear as a single figure'. We have a situation then where the object of medical study is produced via the medical gaze. This is a sexualised process wherein the production of sexually differentiated bodies plays an integral part. The dualistic differentiation that divides all infants into male and female is the result of a long process of the medicalisation of sexual difference within Western medicine and philosophy. Within Western metaphysics dichotomous terms mutually define each other. The sexualised couplet of male and female is a case in point. These two terms are mutually interdependent and mutually productive. In so saying, I underscore the mutability of these categories and a plasticity within the ontological category of sex.

The duality of male and female is based on a sexually defined difference. Sexualised bodies form the basis of this delineation. I wish to state that for man to define himself, through the medicalisation of sexual difference, he constructs woman as his other. Her body provides the basis of this difference. Indeed, her body can be made to speak its truth. In her historical analysis of women and modern science, Londa Schiebinger (1989: 213) cites Samuel Soemmerring, a nineteenth century anatomist, as asserting, 'the anatomist did not have to take a *moral* stand in this matter because the body spoke for itself'. Soemmerring was the first anatomist to construct a comparative anatomy of male and female skeletons. Within his representation of the female's skeletal difference, he exaggerated the breadth of the pelvis.[4] He also enlarged, again in its difference to the female, the size of the male cranium. The female body as a text that speaks her own truth tells, through her embodied difference, her destiny as child-bearer. This predestination is due to her pelvic difference. The male in contrast is destined for rational thought through his anatomical and immutable cranial difference. As Soemmering claims, the locatedness of this immutable difference within the body ensures that these matters are a truth spoken by the body itself. The destiny of male versus female is then a matter of immutable truth and not a moral or ethical issue thus ensuring objectivity. It is through autopsy that she speaks her truth to the male anatomist who is 'seeing for himself'. Thus the corpse inscribed as female serves to define male identity through her readable embodied difference.

The male observer looks into the opened book of the female body via autopsy and finds there his own body represented by its absence and

"The proper study of mankind is man"

Alexander Pope, 1733

In this area, you can undertake your own voyage of discovery – observing, measuring and experimenting on yourself, and coming to see the body through scientific eyes.

Over the centuries, our knowledge of the human body has been built up through careful observation, measurement and experiment.

Gradually, objective study has supplanted myth and a growing understanding of how the body works in health and sickness has led to dramatic changes in the way we perceive, fight and overcome disease.

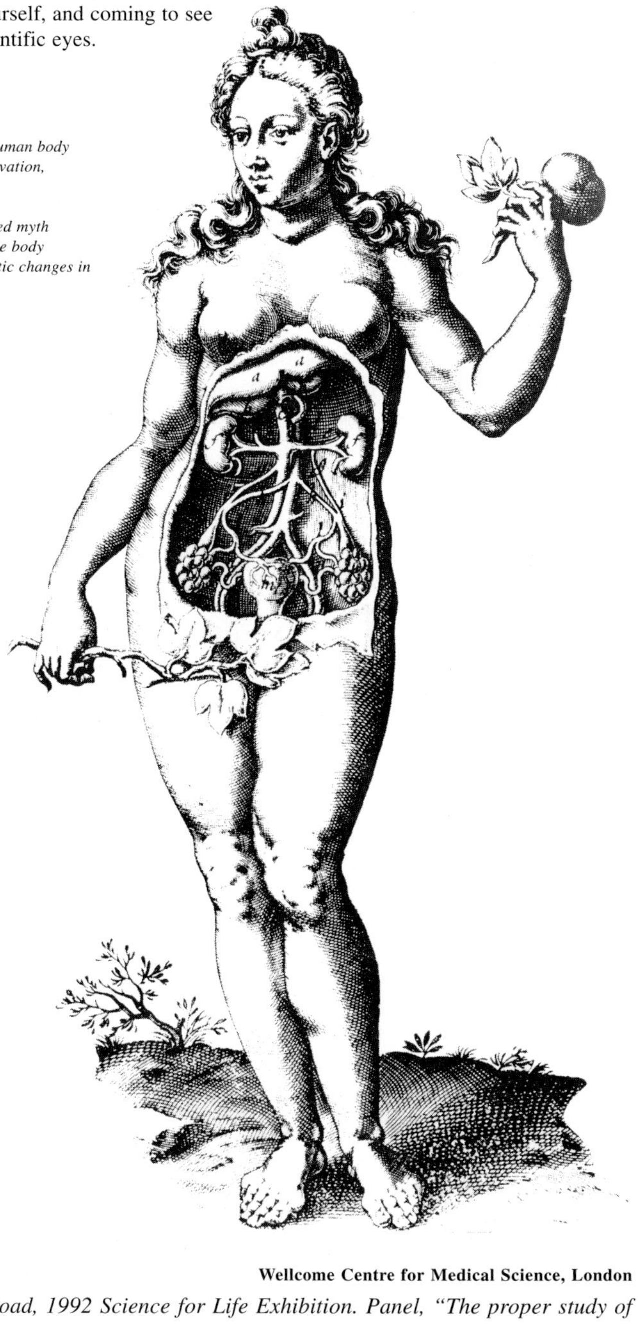

Wellcome Centre for Medical Science, London

Wellcome building, 183 Euston Road, 1992 Science for Life Exhibition. Panel, "The proper study of mankind is man".

difference. In this respect, then, for man to know himself he has medically defined himself in his difference from those he marked as Other. The installation that met the gaze of the visitor to the *Science for Life* exhibition on display at the Wellcome Institute for Medical History in London in early 1994 provides a lovely representation of my argument. This is a juxtaposition of image and text representing the quest for scientific knowledge of the body. The inscription reads 'The proper study of mankind is man'. Juxtaposed with this is the text of his difference, the opened body of a woman. Cynthia Russett (1989: 2–3), in her discussion of Alexander Pope's admonition that 'the proper study of mankind is man', states, 'If the proper study of mankind was indeed man, it was discrete, not universal, man – humanity divided by class, nation and race. And increasingly it seemed, as the century wore on, that the proper study of mankind was woman'. I will return to this idea of identity through marked difference later. I will conclude this section with the thought that it is through his difference to those marked as Other that man comes to know himself.

Medical Gaze as Effecting Embodied Sexual Differentiation

> What determines me at the most profound level, in the visible, is the gaze that is outside.
>
> (Jacques Lacan, 1977: 06)

The medical gaze, then, represents an incredible power to define, classify, categorise and *police* the resultant categories. Foucault (1973: 9) states, '... it was no longer the gaze of any observer but that of a doctor supported and justified by an institution, that of a doctor endowed with the power of decision and intervention'. I want to now look specifically at the medical gaze and sexual difference. I will do this by briefly describing the career of Gottleib Gottlich. I am using Gottlich's story to reinforce the idea that the medical gaze effects embodied sexual difference. It will also become clear that I am using Gottlich's story to introduce the idea of ambiguous sex.

P.D. Handyside, in the Edinburgh Medical and Surgical Journal of 1834, presents Gottleib as a case study to his peers. Handyside (1834: 313–318), reports,

> *Gottlieb Gottlich,* the son of an officer in the Saxon service, was born at the village of Nieder Leuba, near Zittau, in the kingdom of Saxony, on the 6th of March 1798.
>
> He was considered at birth to be of the female sex and was accordingly baptised by the name of Maries Rosine Gottlich; ... He was brought up and entered into service as a female, without his true sex being ever suspected, even by himself, until the spring of 1831, when, on account of a fracture of the left os femoris, he was received into the hospital at Dresden.

Handyside's narrative structure reveals the retrospective insertion of true sex into this tale. At this stage, however, both Gottleib and her family and community unproblematically consider her to be female. Handyside continues,

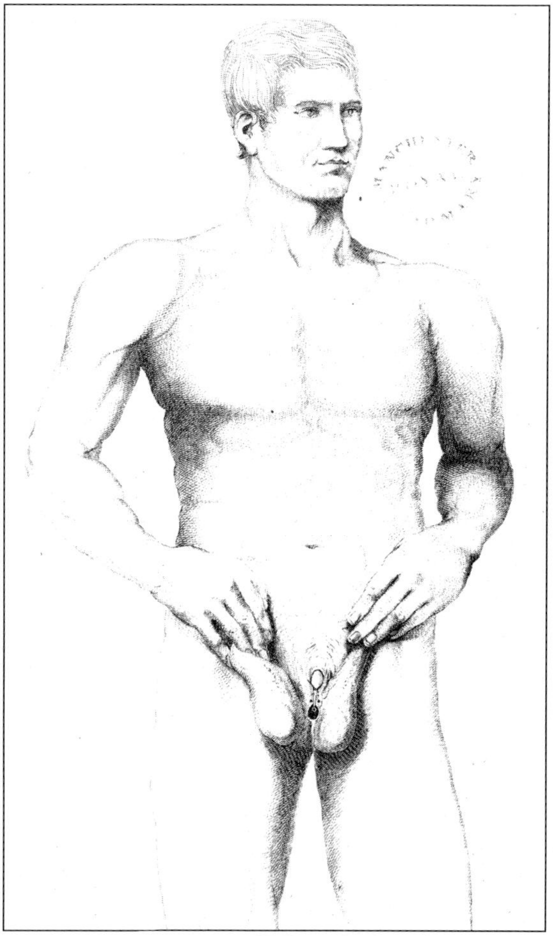

Wellcome Institute Library, London

Engraving: Hermaphrodite.

Plate I, p. 313, vol. xliii in Handyside, P.D. (1835), 'Account of a Case of Hermaphrodism', The Edinburgh Medical and Surgical Journal, *vol. 43.*

> He still continued to clothe himself in female attire and to wear his hair turned up, until November 1832, when he visited the University of Heidelberg; and then it was that I first had the opportunity of seeing him. Professor Tiedemann, before whom he presented himself, did not hesitate to regard him at once as a male, but declared him to be (in the words of that distinguished anatomist) 'evidently a man with genitals of uncommon conformation. She will dress herself, therefore in men's clothes and adopt the name of Gottleib'. Professor Blumenbach of Gottingen, to whom this individual was presented on the 4th February 1833, confirmed the view of Professor Tiedemann.

When Gottleib is seen by three representatives of the medical profession her body comes under the gaze and she is unhesi-tatingly pronounced to be 'he'. He is named Gottleib by the illustrious Professor and brought into being as male. Gottleib's body has undergone a change in gender from female to male. Accompanying his sex determination he is renamed and reclothed via the medical gaze. According to Handyside's narrative:

> Since that period Gottlich has been employed exhibiting himself to the members of the medical profession at the different schools in various countries in Europe. After visiting Bonn, Jena, Marburg, Mainz, Offenbach, Breslau, Bremen and other places on the continent, he travelled by way of Hamburg and Cristiania to London in April last, where he was examined by Sir A. Cooper, Mr Green, Dr Davis and Dr Grant. He next exhibited himself at Manchester, Liverpool, Cork, Dublin, Glasgow, Aberdeen and Montrose and arrived at Edinburgh in November last.

Not only has Gottleib been named by the medical profession his identity is to become chimeric. He is deemed to be neither male nor female but perhaps an hermaphrodite. After five pages of detailed anatomical description Handyside, however, states,

> Under the impression then, that such is the just and correct anatomico-physiological definition of an hermaphrodite and there

> is much reason to believe that no strict instance of such a condition of the generative organs – a union of all the organs of both sexes in the same individual – has ever occurred either in the more perfect animals, or in the human species, I would designate the forgoing case of Gottlich as one simply of Hypospadius, ... which is only the first step toward a true and real hermaphroditism.

As pre-hermaphroditic, Gottleib's condition was deemed to be not 'beyond the reach of surgical art'. Handyside pontificates thus,

> The present case is one in which we might fairly anticipate success from such as operation. When it was formerly proposed, however, by Professor Langenbeck, Gottlich declined all surgical aid and he is now still more adverse to a proposal of this kind, since it would at once deprive him of his present easy and profitable mode of subsistence. March 1835, pp. 313–318.

Gottlich had been transformed from a young woman in service to a touring object of the gaze. His identity had been subsumed into his role as exhibit. This role is one deemed by Handyside to be both easy and profitable. This is in contrast to a life neither easy nor profitable 'in service'.

Gottleib Gottlich's story, as told by Handyside, illustrates the power of the medical profession to name and construct the criteria within which masculinity, femininity and their Other, the chimera of hermaphroditism, are named. In this respect, hermaphroditism seems to function as other to a dualistic notion of sexual differentiation that contains both the male and female. Hermaphroditism also seems to be that which falls outside the categorisation in order to anchor it there as originary. The monstrous deformed and deviant connotations serve to further reinscribe dualistic and separate notions of male and female as the norm.

Lives Lived and Autopsies Performed

The nineteenth century was the time identified by Foucault when it became necessary to have a true sex. Hermaphrodites were renamed as pseudo-hermaphrodites and the truth of their real sex was vigorously sought. The 'true hermaphrodite' became a not-being, who occupied a place only in the imaginary sphere of the chimeric and monstrous. Foucault (1980: viii–ix) in the introduction to *Herculine Barbin* states,

> From the medical point of view, this meant, when confronted with an hermaphrodite, the doctor was no longer concerned with recognising the presence of the two sexes, juxtaposed or intermingled, or with knowing which of the two prevailed over the other, but rather with deciphering the true sex that was hidden beneath ambiguous appearances. He had, as it were, to strip the body of its anatomical deceptions and discover the one true sex behind organs that might have put on the forms of the opposite sex.

The necessity to strip the body of its anatomical deceptions in order to discover the true sex, makes the autopsy the ideal forum within which to make this discovery. It is in the depths of the body that the truth can be detected. The external body in its representation of the true sex via the genitals can be deceptive.

In this respect, the procedures of dissection and autopsy are central to the development of a medicalised sexual differentiation. As I have indicated earlier, the body of woman functions in her difference to that of man. There is a necessity in this relationship, the terms serve to define each other in and through their difference. Attendant to this is the abject difference of the hermaphroditic body that functions as Other to the heterosexualised duality of male and female. I will briefly return to the notion of the hermaphrodite as abject, but before I do that let us return to the autopsy theatre.

Ruth Richardson (1987: 1) in *Death, Dissection and the Destitute* identifies the nineteenth century as a time when the corpse became an anatomical object. She states, '... by the turn of the nineteenth century, both the dead and the living body had alike become objectified or "reified" within the anatomical and surgical fraternity'. Richardson's analysis is commensurate with the Foucauldian assertion that at this time the human body was produced as an object of the medical gaze. It is also commensurate with my assertion that the human body, and in particular the corpse, became the primary text of anatomy. In autopsy, the body is viewed in its depth. Michael Mahon (1992: 3) states, 'In autopsy the invisible depths of the individual body can be brought to light; it enables what was once merely a surface gaze to penetrate to the hidden interior'. The construction of the body as a text of difference extends through corporeal materiality. Andrew Fyfe's kindly admonition that, 'The scalpel should be held somewhat after the manner of a pen', beautifully illustrates the idea of body as text. But it is as a text not only to be read but also to be written. In other words, the procedure of autopsy facilitates the *inscription* of body as text.

I want to now look at three examples of autopsies in order to ascertain the truth of sexual difference. These are the Chevalier d'Eon, Herculine Barbin and James Carey stories. Within the story of the Chevalier D'Eon's life and subsequent death resides a sex that is defined in legal terms during life and subsequently by the power of medicine post-mortem. In his preface to White's translation of the Chevalier's memoirs, Gaillardet (1970: vii) comments, 'After having given witnesses tangible proof of his first sex, he provided other witnesses tangible proof of his second'. It was indeed as if, 'two persons were mysteriously embodied in one human being'. This enigmatic status of the Chevalier's sex could only be solved on the autopsy table. The 'impenetrable veil' of the Chevalier's sex was lifted 'completely and forever' following, 'the inquest conducted on his sexual organs after his corpse had been dissected in the presence of the most competent authorities' (Gaillardet, 1970: ix).

The Chevalier d'Eon's sex was legally defined during his life. This legal intervention came about in response to a wager made concerning the sex of this chimerical individual. In the 1865 edition of his *Principles and Practice of Medical Jurisprudence,* Alfred S. Taylor discusses this legal case. In the late 1700s in London there occurred a 'craze of betting' regarding the sex of the Chevalier. Reputedly £120.00 had been underwritten (C.J.S. Thompson, 1938: 192). The enigmatic Chevalier declined to reveal anything, resulting in a legal action being taken by Mr Hayes (a surgeon) to recover 700 guineas he had lodged with a broker. On 4 October 1771 the case was tried before Lord Mansfield. The jury, acting on the testimony of a surgeon who had attended the Chevalier, deemed her to be a woman. This legal definition of the true sex of the Chevalier did not allay doubt and speculation. Indeed, upon his or her death the perfect opportunity arose to clarify the situation once and for all. An autopsy carried out on the corpse would solve once and for all the mystery. Let us turn now to the story as told by White (1970: 311).

> The Chevalier d'Eon died in London at 26 New Milman Street, on the 21st May, 1810 at the age of eighty three. At the news of his death Queen Charlotte tried to ensure that the secret of his sex was buried with him. But the men she sent to guard the body could not restrain the flood of curiosity his death had aroused in high places. In spite of all the measures she had taken to prevent it, the corpse was minutely examined by surgeons in the presence of eminent witnesses and subjected to a post-mortem which established beyond doubt that it was that of a man.

Through the procedure of autopsy the truth of the Chevalier's sex can be read from his corpse where is it fixed after death. Foucault, while introducing the reader to the memoirs of Herculine Barbin, tells us of a story by Oscar Panizza entitled *A Scandal at the Convent*. Panizza fictionalised Barbin's life as the basis for his novel. He, however, substituted for Barbin a shadowy figure with no fixed identity. Foucault (1980: xvi) comments on the chimeric nature of her identity, 'He did not even choose to fix her with a suicide, whereby she would become a corpse, like Abel Barbin, to which curious doctors in the end assigned the reality of an inadequate sex'. The character from Panizza's text vanishes, 'at the end of the narrative leaving no trace'. In contrast, the truth of Herculine Barbin's true sex, like the Chevalier's, was fixed by the medical gaze within the objectified corpse she became. In this respect the corpse is something different to the person who occupied the body in life. The corpse is an inanimate, unoccupied, inert object of scientific inquiry. The corpse is produced through medical and scientific discourse to become the object of that subject. The corpse as anatomical object then belongs to the emerging discipline of anatomy. In other words, the corpse as anatomical text belongs to the discourse through which it is inscribed. The hermaphroditic corpse as anatomical object par excellence by necessity belongs to the discipline who produced its specificity. There is an inevitability here that corpses marked as different will be read as objects of difference by the very subject that created them as such.

Herculine Barbin knew of the inevitability of the autopsy that would be performed upon her corpse. She writes, 'When that day comes a few doctors will make a little stir around my corpse; they will shatter all the extinct mechanisms of its impulses, will draw new information from it, will analyse all the mysterious sufferings that were heaped up on a single human being' (as cited by Foucault, 1980: 103). E. Goujon was to perform the autopsy assisted by Doctor Reigner. Goujon writes, 'The observation that I am reporting here is certainly one of the most complete that science possesses of this kind, since the individual who is the object of it could be followed, so to speak, from his birth until his death and since the examination of his corpse, as well as the autopsy, could be performed with all the care to be desired' (as cited by Foucault, 1980: 129–130).

Goujon refers to an autopsy performed on another corpse marked as hermaphroditic. He states, 'The autopsy permitted a rectification of the first judgement which had determined his sexual identity during the greater part of his life and a confirmation of the exactness of the diagnosis that in the end had assigned him to his true place in society' (as cited in Foucault, 1980: 131). It is the place that one occupies in society that seems important here. Autopsy can determine whether a position in society was occupied fraudulently or not. It appears that the ability to occupy a sexed position in society is determined by one's sexualised anatomical difference. Again here anatomy determines destiny. Civil status is read from the text of the body.

Like Herculine Barbin, James Carey knew of the inevitable autopsy that would be performed on his corpse. The excerpt with which I opened this chapter comes from a pamphlet published in 1839 verbosely called *Facts connected with the life of James Carey whose eccentrick habits caused a post mortem examination by Gentlemen of the Faculty to determine whether he was Hermaphroditic: with Lithographed Drawings, made at their request.* The first page of the pamphlet is prefaced with the following ditty ...

> Facts reveal'd by Goddard's knife,
> Sheds light upon the M.D. strife;
> (For centuries contended.)
> That nature constant in her plan,
> Immutable in forming man,
> The sexes pure intended.
>
> But Carey's form, outre and strange!
> Discloses a most awful change;
> Where virile organs dwell.
> Defects of ducts, of glands and muscle
> Brains of Physiologists puzzle,
> To solve the mystic spell!

Within this ditty lies a wealth of information regarding the immutability of heterosexualised sexual difference and the simultaneous production and erasure of the hermaphroditic body as its abjected other. We can ascertain that Nature was immutable in forming man. Man could be determined, in his difference, to woman. This difference is determined by presence or absence of the virile organ. The purity and immutability of the difference between man and woman is intended within the plan of nature. For centuries the strangeness of ambiguous bodies, such has Carey's has caused consternation and puzzlement. Following death, however, the surgeon's knife can open the book of Carey's body and from its depths read the truth, in order to solve the mystery of the mystic spell.

Again the hermaphroditic corpse awaits reading by the authority of the medical gaze. Thirteen medical doctors examined Carey's opened body under autopsy on 5 June 1838. His corpse was dissected and, like that of Barbin, the parts deemed to have a bearing on the hermaphroditic difference were excised. The plates within the pamphlet fix, within the text, these 'anatomic anomalies' for the public gaze.

Hermaphroditic Bodies as Abjected Bodies

In the above three cases autopsy is an inevitable necessity for the fixation of a dualistic sexual difference. When bodies that challenge, contest, or subvert the dualistic nature of sexual determination come under the medical gaze, their 'true' sex is determined and assigned. The 'natural fact' of dualistic sexual differentiation is thus *produced* and held in place by the abject chimera of hermaphroditism. Foucault (1976: 38) states, 'their very being, confounded the law that distinguished the sexes and prescribed their union'. Hermaphroditism therefore must take up the status of not-being in order to provide the foundation upon which heterosexualised sexual difference stands. With the hermaphrodite assigned the location of not being it is essential that people whose bodies are ambiguously sexed have a 'true sex' discovered and assigned even if this has to be done post-mortem. This necessity serves to reinscribe a dualistic sexual difference that is always already in existence.

In this respect, the hermaphroditic body comes into existence only to be simultaneously denied existence. It is the ambiguous nature of the hermaphroditic body that threatens the law that holds the dualistically defined sexes apart. It is also the hermaphroditic body as this couplet's abjected other that simultaneously holds them in place.

> This boy–girl, this never eternal masculine–feminine, is nothing more than what passes at night in the dreams, the desires and the fears of everyone.
>
> (Foucault, 1980: xvi)

Notes

1. Much of the research for this chapter was conducted at the Wellcome Institute of Medical History in London. I would like to thank the staff there for their assistance. I would also like to thank the Waikato branch of the New Zealand Federation of University Women for their provision of the Merit Award which provided much needed and appreciated funding for this exercise. I also appreciate the comments from Dr Radhika Mohanram regarding an earlier draft of this chapter. The material regarding James Carey would not have been accessible to me without the excellent assistance of Charles Greifenstein the Historical Reference Librarian at The College of Physicians of Philadelphia.
2. In *Instincts and their Vicissitudes,* Freud (1957: 134) speaks of the necessity for living organisms to distinguish between an 'outside' and an 'inside'. For the subject to become a subject it must make the differentiation between Subject (ego) and Object (external world). Freud correlates this differentiation with that of active and passive. This is, however, not to be conflated as, 'The relation of the ego to the external world is passive in so far as it receives stimuli from it and active when it reacts to these' (1957: 134).
3. Judith Butler (1990) within her text *Gender Trouble* argues that sex and gender are both constructed through repetitive performative acts. She develops her argument regarding the mutability of corporeality within *Bodies that Matter*. Elizabeth Grosz (1994) also deals with the body as effect within her text *Volatile Bodies.* For further discussion of corporeal mutability see also Vicky Kirby's (1991) 'Corporeal Habits Addressing Essentialism Differently'.
4. See Schiebinger's fascinating discussion of this in Chapter Seven of her *The Mind Has No Sex?*

References

Akin, J. (1839), *Facts Connected with the Life of James Carey whose Eccentrick Habits caused a Post Mortem Examination by Gentlemen of the Faculty to Determine Whether he was Hermaphroditic: with Lithographed Drawings made at their Request*, Philadelphia.

Anon (1831), *The Quizzical Gazette; or Moral and Satirical Reflector*, Saturday, 10 December.

Anon (1902), 'Hermaphroditism: Sex Detected Post Mortem', in *The British Medical Journal*, 8 November, p. 71.

Bell, B. (1916), *The Sex Complex*, London: Balliere, Tindal and Cox.

Butler, J. (1990), *Gender Trouble: Feminism and the Subversion of Identity*, London: Routledge.

Butler, J. (1993), *Bodies that Matter; On the Discursive Limits of 'Sex'*, New York: Routledge.

Conrad, L.I. (1995), 'The Arab-Islamic Medical Tradition', in Conrad, L. *et al.*, *The Western Medical Tradition 800BC to AD1800*, Cambridge: Cambridge University Press.

Foucault, M. (1970), *The Order of Things; An Archaeology of the Human Sciences*, London: Routledge.

Foucault, M. (1973), *The Birth of the Clinic; An Archaeology of Medical Perception* (trans. Sheridan, A.), New York: Routledge.

Foucault, M. (1976), *The History of Sexuality. Volume One An Introduction* (trans. Hurley, R.), London: Penguin.

Foucault, M. (1977), *Discipline and Punish; The Birth of the Prison* (trans. Sheridan, A.), London: Penguin.

Foucault, M. (1980), *Herculine Barbin; Being the Recently Discovered Memoirs of a Nineteenth-Century French Hermaphrodite* (trans. McDougall, R.), New York: Pantheon Books.

Freud, S. (1957), 'Instincts and their Vicissitudes', in *The Standard Edition of the Complete Psychological Works of Sigmund Freud* (trans. Strachey, J.), London: The Hogarth Press.

Fyfe, A. (1826), *A Compendium of Anatomy Human and Comparative; with Directions for Dissecting the Different Parts of the Human Body*, Edinburgh: Adam Black.

Gaillardet, T. (1970), 'Preface', in White, A. (trans), *The Memoirs of the Chevalier D'Eon*, London: Anthony Blond.

Gilman, S. (1989), *Sexuality An Illustrated History*, New York: Wiley.

Grosz, E. (1994), *Volatile Bodies*, Sydney: Allen and Unwin.

Handyside, P.D. (1835), 'Account of a Case of Hermaphroditism', in *Edinburgh Medical and Surgical Journal*, pp. 313–318.

Jordanova, L. (1989), *Sexual Visions; Images of Gender in Science and Medicine between the Eighteenth and Twentieth Centuries*, New York: Harvester Wheatsheaf.

Kirby, V. (1991), 'Corporeal Habits: Addressing Essentialism Differently', *Hypatia*, vol. 6, no. 3, Fall.

Knight, B. (1980), *Discovering the Human Body: How Pioneers of Medicine Solved the Mysteries of Anatomy and Physiology*, London: Heinemann.

Lacan, J. (1977), *The Four Fundamental Concepts of Psychoanalysis* (trans. Sheridan, A.), London: The Hogarth Press.

Laqueur, T. (1990), *Making Sex; Body and Gender from the Greeks to Freud*, Massachusetts: Harvard University Press.

Mahon, M. (1992), *Foucault's Nietzschean Genealogy. Truth, Power and the Subject*, Albany: State University of New York Press.

Money, J. in Cole, J.K. and Dienstbier, R. (eds) (1973), *Nebraska Symposium on Nature*, Nebraska: Nebraska University Press.

Neve, M. (1995), 'Conclusion', in Conrad, L. *et al.*, *The Western Medical Tradition 800BC to AD 1800*, Cambridge: Cambridge University Press.

Nutton, V. (1995b), 'Roman Medicine, 250 BC to AD 200', in Conrad, L. *et al.*, *The Western Medical Tradition 800BC to AD 1800*, Cambridge: Cambridge University Press.

Nutton, V. (1995c), 'Medicine in Late Antiquity and the Early Middle Ages', in Conrad, L. *et al.*, *The Western Medical Tradition 800BC to AD 1800*, Cambridge: Cambridge University Press.

Nutton, V. (1995d), 'Medicine in Medieval Western Europe, 1000–1500', in Conrad, L. *et al.*, *The Western Medical Tradition 800BC to AD 1800*, Cambridge: Cambridge University Press.

Richardson, R. (1987), *Death, Dissection and the Destitute*, London: Routledge.

Riley, D. (1988), *Am I that Name? Feminism and the Category of 'Women' in History*, London: Macmillan Press.

Russett, C.E. (1989), *Sexual Science: The Victorian Construction of Womanhood*, Cambridge, M.A.: Harvard University Press.

Schiebinger, L. (1989), *The Mind Has No Sex? Women in the Origins of Modern Science*, Massachusetts: Harvard University Press.

Singer, C. (1926), *From Magic to Science; Essays on the Scientific Twilight*, London: Ernest Benn.

Singer, C. (1957), *A Short History of Anatomy and Physiology from the Greeks to Harvey*, New York: Dover.

Taylor, A.S. (1865), *The Principles and Practice of Medical Jurisprudence*, London: John Churchill and Sons.

Thompson, C.J.S. (1938), *Mysteries of Sex: Women who Posed as Men and Men who Impersonated Women.* London: Hutchison and Co.

Wear, A. (1995), 'Medicine in Early Modern Europe, 1500–1700', in Conrad, L. *et al.*, *The Western Medical Tradition 800BC to AD 1800*, Cambridge: Cambridge University Press.

White, A. (trans.) (1970), *The Memoirs of the Chevalier D'Eon*, London: Anthony Blond.

10

The Imperfect Woman

The Female Figure in Dürer's *Melencolia I*

Heleen Buijs

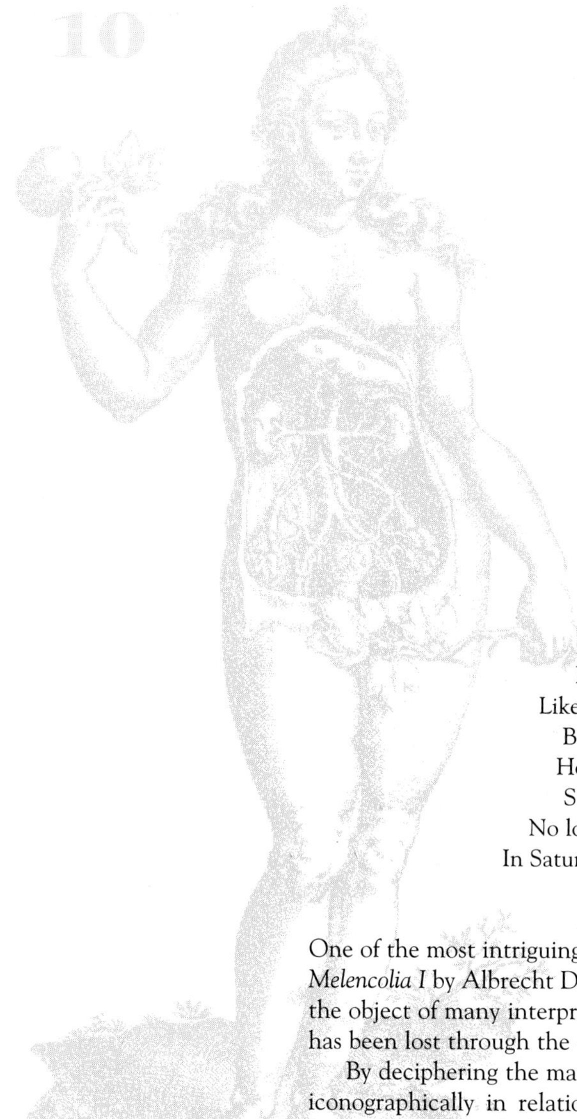

God has given me unduly
In my nature melancholy.
Like the earth both cold and dry,
Black of skin with gait awry,
Hostile, mean, ambitious, sly,
Sullen, crafty, false and shy.
No love for fame or woman have I;
In Saturn and autumn the fault doth lie.
(Anonymous)[1]

One of the most intriguing works in the history of Western art is the etching *Melencolia I* by Albrecht Dürer (1471–1528). Dated 1514, the print has been the object of many interpretations as its original meaning given by the artist has been lost through the centuries (Figure 10.1)

By deciphering the many elements in the print and by interpreting them iconographically in relation to Dürer's life and works,[2] to his interest in humanistic art and philosophy, astrology and alchemy,[3] art historians have tried to identify the meaning of this print. Most researchers have given interpretations of Melancholy's facial expression, of her being winged and of her general appearance, identified her as an allegory of melancholy and considered the matter closed. The presence of Melancholy as a woman, however, they have taken for granted.[4] Yet we may assume that Dürer has given a special meaning to the presence of a woman in his print, as he has given all elements a specific meaning. Along with her bodily appearance and her facial expression, her presence as a woman needs to be taken into consideration in the reading of this print, especially because she is unlike the women in other works by Dürer. This chapter offers an interpretation of why Melancholy is a woman.

The etching shows a robust woman sitting on a pavement. Her left elbow is leaning on her knee, the hand clenched to a fist, supporting the head.

Figure 10.1: Melencolia I

Holding a compass in her right hand without using it, she rests her right arm on a closed book. Her rich dress is that of the bourgeois woman; keys are dangling from a ribbon attached to the belt and her purse has fallen on the ground. Her face is in shadow, but her eyes are lightened up in the dark face. The tips of two large wings are visible from behind her back, touching the hourglass and the 'magic' square respectively. Next to her sits a little putto, also winged, on a millstone that is decorated with a cloth. The putto is very busy scribbling something on a slate. A lean dog is sleeping at their feet. Around this trio objects have been placed such as scales, an hourglass, a bell, a cube, a sphere and a plane. In the left upper corner a landscape is depicted in bright sunlight. Dürer has made it seemingly easy for us to find out what the meaning of this print could be: a banner held by a flying bat contains the name *Melencolia I*.

However, the title does not clarify the presence of elements like the instruments that are not related to the traditional depiction of melancholy. For a better understanding, an analysis of the history of the representation of melancholy is needed. Because it is still the most prominent interpretation among art historians, I will discuss Panovsky's perspective.[5] Later Klibansky, Panovsky and Saxl made this view more profound but, like their colleagues, they have paid scarcely any attention to the fact that the central figure of the image is a woman.

The Four Humours

Dürer's concept of melancholy had its roots in antiquity. The word itself derives from the Greek term for black bile. Until the seventeenth century, it was widely believed that black bile together with phlegm, yellow bile and blood formed the four constituent elements of the human body.

Moreover, according to this framework of humoral pathology, the four humours had to be held in equilibrium. Health depended on the four fluids being in balance. A person would become sick when one of the fluids had a predominance over the other three and this sickness was expressed in terms of too hot, too cold, too moist or too dry. Curing the sick was only possible by influencing the relation of these four fluids, for instance by giving the patient wine to make him or her moister, or by bloodletting to expel too much moisture.

In Antiquity and the early Middle Ages physicians and philosophers did not perceive body and mind as distinct from each other; there was a direct link between the four humours and temperament. Someone who had too much blood had not only a hot and moist body but also an over-excited mind. With respect to melancholy, a person was thought to have too much black bile in the body and therefore to be too cold and too dry. The melancholic suffered from a bodily disease and at the same time from a sombre mood. This state was, however, viewed fairly neutrally equal to the other bodily fluids and only inducing illness when the balance between the fluids had been disturbed.

The concept of the four humours was embedded in an overall cosmology in which all elements in macro cosmos and micro cosmos had found their place. In this way the Greek philosophers were able to give a simple, understandable solution to the complexity of the world. It was held that four elements – fire, water, air and earth – formed the creative basis of all that existed. Those four elements corresponded with the four humours, with the four 'ages of man' and with the four seasons, the latter two representing the divisions of time. All these aspects formed a coherent system and they were believed to influence each other. The nature and extent of that influence could vary, for instance according to one's age. Melancholy was ruled by the black bile, that was connected to the earth, to autumn, to the evening and to 'manhood' – the age of the four 'ages of man' which was between 40 and 60 years. Someone of an older age was understood to be more sensitive to the influence of autumn and of the planet Saturn and was more likely to have a melancholic mood than was a younger person. Everyone's character, talent, fate and even moral characteristics were determined in this cosmology.

Humoral pathology was common until the seventeenth century when medical research became rationalised as a non-mythological concept with a scientific base. While it underwent only a few changes, it did shift towards a different interpretation of the doctrine of the four humours and especially of the early medieval concept of melancholy. Melancholy became ever more endowed with unfavourable properties both of body and mind. Some of these negative characteristics came from treatises on mental illnesses (Klibansky, Panovsky and Saxl, 1964: 291). It became medical belief that extreme forms of melancholy could bring a person to insanity. Medical scientists started considering melancholy as worse than the others' state of mind. With negative bodily aspects like black hair and swarthy skin, the melancholic was 'awkward, miserly, spiteful, greedy, malicious, cowardly, faithless, irreverent and drowsy, surly, sad, forgetful, lazy and sluggish; he shuns the company of his fellow-men and despises the opposite sex' (Panofsky, 1943: 158). The four humours were ordered hierarchically, with the sanguinic first, then the choleric, and the phlegmatic and melancholy ranked lowest.

Parallel to this negative conception of melancholy as a disturbed mental and physical state ran a positive conception which went back to Aristotle (fourth century BC). He had posed the question: 'Why is it that all those who have become eminent in philosophy or politics or poetry or the arts are clearly melancholic and some of them to such an extent as to be affected by diseases caused by the black bile?' By analysing this problem Aristotle distinguished between two types of melancholic men: the normal man whose bodily fluids are in equilibrium and who can becomes melancholic because of a temporary preponderance of the black bile; and the 'natural melancholic', mentally and bodily different from normal, who has an abnormal intelligence. With this answer, the concept of the melancholic as a genius was introduced. Melancholy was not an illness but a permanent mental and bodily state. In this conception, the melancholic genius did not possess the same negative characteristics as the melancholic as perceived in humoral pathology. Medieval philosopher Albertus Magnus makes this distinction clear: 'Melancholics are thin, short and swarthy. But those who are of that sort of melancholy which is warm and affected by "adustio" [that is to say, the Aristotelian great men] are very tall and slender and dark and have firm flesh'.[6] With the neo-platonic revival in the Renaissance, the melancholic genius became an object of interest in medical and philosophical thinking to a greater extent than it had been during the Middle Ages. Klibansky *et al.*, have ascribed the reassessment of the concept of the melancholic genius to Ficino (1964: 254–274). In *De vita Triplici* he identified men such as Plato, Aristotle and Michelangelo as melancholic geniuses and observed that: 'it was the black bile that obliges thought to penetrate and explore the centre of its objects and it raises thought to the comprehension of the highest'. This relation between the black bile and the genius man had yet two sides. On one hand, a particular proportion of the black bile in relation to the other humours could create a 'natural' melancholic and could enable that man to have brilliant thoughts. On the other hand, hard contemplation could temporarily bring a healthy man into a melancholic mood. By thinking hard and deeply man was using the fluids of the mind and when he had used them up, he would become dry and cold and, therefore, melancholic.

Melancholy in Representation

Representations of the four humours can be found throughout the history of Western art, predominantly in representations of the Four Temperaments. Images of these Four Temperaments depict the four humours combined with references to the four seasons, the four ages of man and the four elements. In this purely descriptive representation each temperament is typified by a single figure, more or less inactive and distinguished mainly by physique, expression, costumes and attributes (Figure 10.2)

Figure 10.2: The Four Temperaments

Melancholy has been depicted as a person supporting the head with a hand, a phenomenon found as far back as antiquity in depictions of Saturn, the god who ruled melancholy. As humoral pathology increasingly endowed it with negative connotations, Melancholy became visualised with these specific characteristics. The miserly nature of the melancholic was represented in the purse he is holding in his right hand and the money chest that his right elbow is resting on.

The melancholic genius, however, has mostly been represented as a man lying on a bed with the head supported by the hand and accompanied by other figures. Medieval statesmen, poets, evangelists and Church Fathers were considered to be of this category (Panofsky, 1943: 287).

Figure 10.3: Melancholics

There is a third category of representations of the four temperaments: depictions of couples, mostly a man and a woman, who enact a scene typical of their particular temperament. These scenes derived from depictions of moral human behaviour from the High Middle Ages, which often show 'Acedia' (a turning away from God by laziness and one of the deadly sins in medieval Christian terms) as a sleeping man or a woman with their head resting on a hand (Panofsky, 1943: 159). Subsequently, this image transformed to the representation of Melancholy as a woman fallen asleep over her distaff and a man in the background also asleep, generally at a table (Klibansky *et al.*, 1964: 300–301) (Figure 10.3) The closed book on the man's lap was a sign that no wisdom could be attained by study. Only by doing good in Christian terms and by experience could one become wise.

In Ancient Greek and early medieval times, medical and philosophical thinking about the four humours did not include woman as an object of research. However, one case was mentioned in a manuscript of the second century AD concerning melancholy due to love sickness (Wack, 1990: 9). The medical scientist Galen of Pergamon outlined the case of a dejected and sleepless woman of high nobility. He carefully examined her, questioned her while checking her pulse and eventually determined that her melancholy was due to her being in love with a dancer.

Apart from this one recorded case, it was apparently not until the twelfth century that woman came into the focus of medical thinking. Influenced by Christian moralising tendencies in the Middle Ages, William of Conches (ca. 1080–1154) developed a cosmology in accordance with Christian belief and its moral values. He believed man's ideal state was sanguinic but that it was lost after the Fall and other qualities increased or diminished. Man degenerated into cholerics, phlegmatics and melancholics; the latter was the most unfavourable (Klibansky *et al.*, 1964: 105). Drawing on Genesis to extend the theological basis on which he developed his cosmology, William of Conches argued that as woman was created out of Adam's rib, she was not equal to man: he was made out of earth whereas she came from flesh. William of Conches considered woman colder and moister, more phlegmatic than man who was by nature more sanguine. Therefore, he said, men should only be compared with men, women with women.

The increasing attention to woman is evident in depictions of female figures representing the four 'ages of man'. One of the first writers who distinguished between the female and male type of melancholy was Hildegard von Bingen (1098–1179). Living in Germany, she was abbess of a Benedictine monastery and a scientist who wrote about religious, medical, biological and other subjects. She explained her cosmology, which combined Greek cosmology and Christian belief, in *Causae et Curae.* Like William of Conches, Hildegard considered melancholy a result of the Fall of man: '... when Adam broke the law, the sparkle of innocence was dulled in him and his eyes, which had formerly beheld heaven, were blinded and his gall was changed to bitterness and his melancholy to blackness'.[7]

Hildegard was not afraid to take up an independent position in theological and medical matters. Making a clear distinction between male and female characteristics, she accepted the conventional wisdom that woman was colder and moister, more tearful and complaining than man. But, diverging from the norm that man was more airy (attributed to the sanguinic) and woman more earthy (tending to melancholy), she ascribed the earthy complexion to man, on account of Adam's formation from the earth, and the airy temperament to woman (Newman, 1987: 128).

By presenting separate character sketches for man and woman, Hildegard was the first who offered a discussion of individual types within her generic description of the four humours (Newman, 1987: 132). The sanguine woman had soft, delicate flesh and slender veins. Her complexion was fresh, she was fertile, tender in lovemaking and subtle in arts. Hildegard considered the melancholic woman a complicated type who easily succumbed to sickness and exhaustion. She was better off without a husband. This type of woman found little pleasure in sex, was gaunt in body and wandering in her thoughts.

She was a sort of neurotic intellectual. In these character sketches Hildegard seemed to have worked less from classical theory than from her own observations of the women around her. Her experience as an adviser and confidante to laywomen might have helped her in developing her ideas (Flanagan, 1991: 99).

Especially in thirteenth and fourteenth centuries, representations of lovesick women were related to melancholy and illness. Nonetheless, throughout the Middle Ages man was supposed to be more susceptible to lovesickness than woman because he was dryer and dryness retained an impression upon the imagination more readily than moisture did. In this view, women were more frequently sexually stimulated but had less hope in the realisation of their desires. Therefore, women suffered more than men in a quantitative sense but were more easily cured since their mental impression was less deep.

The late fifteenth century saw the transformation of the single female figure of 'Acedia' to the melancholic couple, often depicted as a sluggish housewife with her sleeping husband (see above). From that time, woman became represented in forms of melancholy other than the love melancholy.

Klibansky, Panofsky and Saxl's interpretation of Dürer's *Melencolia I*

Dürer's etching contains elements derived both from the descriptions and representations of the melancholic in humoral pathology and from the representation of the melancholic couple. The hand supporting the head, the keys as a symbol for the money chest, the purse referring to avarice and the swarthy face have been borrowed from the first group. The clenched fist can also be considered an attribute of avarice (Böhme, 1989: 17). The laziness of the sleeping woman and the closed book come from the second group of representations.

Klibansky *et al.* have indicated the image as a 'symbolic realisation of Melancholia' (1964: 304). Her wings set her apart from all other representations of melancholy. They consider this not a typical example of the melancholic temperament, but an expression of the idea: 'Melancholy is like this'.

With respect to the other objects in the print, Klibansky *et al.* have held that the compass and the sphere in the print represent Geometry, the fifth of the liberal Arts (1964: 317). The magical quadrate, the hourglass with the bell and the scales meaning the measuring of Space and Time are also instruments of Geometry. The seven Liberal Arts, the traditional curriculum of secular learning in the Middle Ages, were given visual form in anonymous female figures. The woman can also be interpreted as the allegory of Geometry (Figure 10.4) Translating Dürer's print as the symbolisation of Melancholy and Geometry in one figure, Klibansky *et al.* have interpreted the image as representing man dealing with the 'art of measuring', the art of creating things.

The place of Geometry relates back to Dürer's views on the art of drawing in perspective and the measuring of all objects, skills he had learned on his journeys to Italy. He considered it a lack of artistic quality when an artist was not able to draw perspective; by contrast, by applying the art of measurement in the right way the artist could gain mastery over nature and over his own work. Klibansky *et al.* contend that Geometry refers here not only to the intellectual capacity of the philosophers and studied men but also to the artistic thinker. In Dürer's words, this is: 'always to pour forth something new in his work' and 'every day to have fresh figures of men and their creatures to make and pour out which no one has seen or thought ever before'.[8]

In addition to avarice, the gesture of the clenched fist could symbolise the fanatical concentration of a mind which has truly grasped a problem, but which at the same moment feels incapable either of solving or of dismissing it.

> Winged, yet cowering on the ground, wreathed, yet beclouded by shadows, equipped with the tools of art and science, yet brooding in idleness, she gives the impression of a creative being reduced to despair by an awareness of insurmountable barriers which separate her from a higher realm of thought.
> (Panofsky, 1943: 168)

Figure 10.4: Geometry

Other figures provide contrasting types to the higher realm of thought. The dozing of the skinny dog signifies, according to Klibansky *et al.*, the dull sadness of a creature entirely losing itself in its unconscious comfort or discomfort. The industry of the writing putto signifies the careless equanimity of a being that has only learned the contentment of activity, even when unproductive and does not yet know the torment of thought.

The figure is Melancholy as well because Dürer realised that the art of measurement could never lead to the discovery of absolute beauty. Dürer said: 'I know not how to show any particular measure that approximates to the greatest beauty'. *Melencolia I* is thus seen by Klibansky *et al.* as Dürer's statement of his own limitations as an artist.

Some Critical Remarks

Several months before Dürer finished his print, his mother died. It has been suggested that the print was the expression of his melancholic mood because of this event. The numbers in the magic square could point to her date of death. Panovsky has dismissed this suggestion, stating that Dürer respected his 'poor and pious mother', that he admired her patience during her illness but that 'his real love belonged to his father' (1943: 170–171). Being so well informed about Dürer's work and writings, Panovsky has not given an explanation or any reference for this statement.

He does, however, comment on Dürer's relationship with another woman – his wife. He noted that 'in her youth she was fairly good-looking and harmless, but in her later years she turned into a somewhat forbidding lady' and that 'Agnes Frey thought that the man she had married was a painter in the late medieval sense, an honest craftsman ...; but to her misfortune her husband discovered that art was both a divine gift and an intellectual achievement requiring humanistic learning' (1943: 6–7). Having grown up in the late Middle Ages, she was simply not able to keep up with Dürer's intellectual development from a medieval craftsman into an intellectual, humanist artist. She was resentful because he often left her alone and preferred the company of his learned friends. In short, they had an unhappy marriage.

Panovsky is suggesting here that his wife was hindering Dürer to become the phenomenal artist he is now known to have been, that she wanted him to be a devoted husband with a regular income and that she did not like him to go beyond the limited borders of their social class. Yet in the late Middle Ages and early Renaissance, marriage was an economic transaction arranged between the two fathers in which the future couple had no voice. Completely subordinated to their fathers or their husbands, women received no other education than that which would make them good and pious housewives. When a couple could not have children, it was the wife who was to blame as her duty was to deliver male offspring to continue the family line. In this context, Dürer's marriage does not seem very different from an average late medieval marriage. By marrying Agnes Frey he entered a higher social class and became financially independent and was able to travel around freely. As her husband he was, within certain boundaries, free to treat his wife as he liked. He might have blamed her for giving him no children, but that cannot be detected from his writings.

Panovsky's remarks, written in the twentieth century and without reference to the wider context of medieval marriages, seem to say more about his own attitude towards women than can be gathered from information about Dürer's marriage. By suggesting that Agnes was complaining wrongfully, he established that woman should be silent, respectful and supportive even when she cannot grasp her husband's intellectual level. Moreover, she is not supposed to play an important role in the life of such a great man who 'loved the company of scholars and scientists, associated with bishops, patricians, noblemen and princes' (Panofsky, 1943: 7). Only by being among other great and interesting men could Dürer develop his extraordinary capacities. Women should stay away from such company, because they threaten to destroy the development of this gift.

From Panovsky's notions of woman and her social roles, we can understand why he has not given her a place in the interpretation of *Melencolia I*. Woman is not supposed to play an important role in the expression of the valuable humanistic thoughts by 'this genius' as her place is in the kitchen not in the study. Unable to interpret her place in Dürer's print, Panovsky has left the presence of the woman as such aside and has given her merely the role of the allegory.

Studying Dürer's oeuvre, it becomes clear that he has made drawings and paintings of women as much as of men throughout his career. Woodcuts, etchings and paintings show biblical, mythological and realistic scenes where women appear as frequently as men. He made special studies of the clothing of women for different surroundings, such as the home, the church and dancing. Under influence of Renaissance Italian artists, Dürer was dealing with beauty and the beautiful form of man and woman . His *Four Books on Human Porportions* (begun in 1500, published in 1528) demonstrate this interest: on his figures of men and women in several sizes (fat, tall, thin, short), he put the exact measurements at various places of the body to ensure he had the correct proportions.[9] In his so-called *Dresdener Sketch Book* he made drawings of female and male figures, hands, feet and children and geometric drawings of buildings, especially in 1512 and 1513 (Musper, 1952: 107). In 1513, he changed to a more personal style in which he succeeded to create the 'ideal' female forms through experience rather than through exact numerical proportions. In this respect we may consider many of his later images of women to be the results of his study of the female figure in the earlier period, particularly between 1500 and 1513. This may clarify why some of the figures seem somewhat awkward before he acquired the skills to draw out of memory and with more intuition. Thus, we may suppose that the representation of the female figure was of importance to Dürer and we have to take this into account analysing the *Melencolia I*.

The interpretation of the etching may also be assisted by considering how women were affected by developments in Dürer's time. This was a period of transition in which medieval notions in religion and philosophy were changing, and in Würtemberg – a town not far from Dürer's home – Luther nailed his proclamations on the doors of the cathedral. In the Middle Ages, leading up to this period, the distinction between the sexes relates back to the doctrine of the four humours and to Aristotelian philosophy. As discussed above, woman was thought to be colder and moister than man. She was less developed than man because lack of heat meant her sexual organs remained internal and unable to concoct perfect semen from blood (Maclean, 1980: 31). One implication of this way of thinking was that man, as the hotter sex, could create more perfect things than woman. It is a duality in keeping with Aristotelian philosophy, in which one element is superior and the other inferior:

> The male principle in nature is associated with active, formative and perfected characteristics, while the female is passive, material and deprived, desiring the male to become complete.
>
> (Maclean, 1980: 8)

Luther's view on woman complies mainly with that in late medieval treatises. In a commentary on Genesis, he described woman as 'a different animal to man, not only having different members, but also being far weaker in intellect'. He compared man with the sun and woman with the moon, the sun being more splendid than the moon. Although both are the most beautiful handiwork of God, because woman is created out of Adam's rib, she does not equal the dignity and glory of the male. In his commentary on Genesis, Luther said that the curse laid upon woman at the Fall makes her suffer from three things: pain in childbirth, sadness and subjection to her husband. Additionally, woman is also more susceptible to avarice, one of three vices, because she is more earthly and therefore, more deeply imbued with it than man (Maclean, 1980: 22).

Neo-Platonism promoted a new theory of beauty and love in the Renaissance. In the visual arts, this new perspective is evident in representations of pagan goddesses, of active, heroic (though not always favourable) females, and the portrayal of virtues as female figures. Such images added to the metaphorical dignity of woman and parallel changes occurred in an economic respect. The position of middle-class women in particular changed to assume an important role in public life in the late Middle Ages and early Renaissance; many, for example were independent tradeswomen (Bornstein, 1978: xi). Renaissance also saw the rise of various writings promoting the superiority of woman above man.

In reaction to these developments, intellectual satire developed which caricatured women, partly to provide a game of amusement for men, partly as the expression of fear of dominance by women (Bornstein, 1978: vi). Simultaneously there remained the medieval notions of woman being less than man, passive, material (opposing form), imperfect and deprived. Widespread, too, were the Lutheran notions of woman as God's beautiful creation who could be compared with the moon but was nevertheless less perfect than man who could be compared with the sun.

These conflicting images of woman in the late Middle Ages and early Renaissance can be traced back in Dürer's *Melencolia I*. The female figure in the print is clothed like a well-to-do bourgeois woman, while her independence is shown by her robustness and central position in the print. The bright eyes in the shadowed face enhance the idea of self-consciousness and intelligence. She can be identified with the new type of tradeswomen – her purse and keys refer to avarice, but also to money business in general – the tradeswoman who represents the woman of a modern era that Dürer was so fascinated by. He has not used the medieval type of melancholy that is thin, spiteful, faithless, sad and lazy; nor has he drawn on the transformation of the sluggish housewife sleeping in her chair as Klibansky *et al.* have suggested. Rather, he has created the representation of a new type of Renaissance woman.

Assuming that the content of the print refers to the Renaissance notion of measuring and the creation of absolute beauty, it could be argued plausibly that he was looking for a female figure representing this new, humanist spirit and that he found her in the independent middle-class woman. All the elements in the print are related to the new ways of thinking. Renaissance aesthetics concerning beauty and the art of measuring are represented in the instruments displayed around the female figure; Lutheran philosophy is reflected in the notion of the sun and the moon. Melancholy is sitting in the dark while moonlight is brightening her eyes that reflect most clearly the mental state she is in. Bright sunlight is shining over the distant landscape. However, Dürer's ambivalence towards the new woman – she was after all threatening male dominance over women – was shown through depicting her as melancholic with negative characteristics, such as the swarthy skin, avarice and passivity. Dürer seemed to be sensible of this aspect of the new woman, as indicated by the less than friendly way he wrote about his wife in his letters.

Melancholy, as the representation of striving for beauty through measuring, could only be a woman because she has all the qualities Dürer needed to express his thoughts in accordance with late medieval and early Renaissance belief. She represents deep thinking as one of God's precious gifts to humankind, like Eve was a precious gift to Adam. However, she possesses less intelligence than man and that explains why she is not able to grasp the highest thoughts and becomes melancholic. An intelligent man cannot fail to understand the most difficult problems in the world, as shown by all the great thinkers from the past to Dürer's present like the humanists and Luther, but a woman can. Melancholy needed to be a woman because she represents imperfection and failure. She is the moon in the night, splendid in her creation of beautiful things but not perfect, while we see the bright sunlight in the distance where all wisdom is. To be there is for her impossible.

Klibansky *et al.* have interpreted the print as an autobiographical statement, as Dürer's recognition of his failure to grasp the complicated requirements of measuring to create beauty. As a northern European artist rooted in medieval thinking and influenced by Luther's theology of humbleness and simplicity, he might have realised that the creation of absolute beauty is in the power of God, not of man. Klibansky *et al.* did not explain the presence of a woman as the central figure in Dürer's print because it appears that they did not want to identify Dürer, the great artist, with a woman. Dürer himself, however, has made the connection: he has given visual form to his recognition of imperfection. Through the creation of an immaterial being he has visualised his extraordinary capacity as an artist which he considered as a precious gift from God. That might explain why Melancholia has been depicted with wings. But, unable to reach the highest standards of perfection, he depicted her as a woman. Melancholy as the imperfect woman is Dürer's recognition of his own limitations.

Notes

1. Quoted in Klibansky *et al.*, 1964: 117.
2. Wölfflin, 1923; Levy, 1964; Braham, 1965.
3. Klibanski *et al.*, 1964; Schuster, 1991; Böhme, 1989; Calvesi, 1993.
4. An exception is Calvesi who associates Melancholy with woman's procreational functions in terms of labour pains and a long pregnancy: Calvesi, 1993: 191.
5. The interpretation of Dürer's Melencolia I has first been put forward by Panovsky, 1943. His ideas underlie the line of thoughts in the book of Klibansky *et al.*, 1964.
6. *De Animalibus Libri*, quoted in: Klibansky *et al.*, 1964: 71.
7. Quoted in: Klibansky *et al.*, 1964: 80.
8. Quoted in: Klibanski *et al.*, 1964: 365.
9. Panovsky, 1915: 84–89. From his analysis, Panovsky reported in this chapter that Dürer must have started drawing the female figure from as early as 1500.

References

Böhme, H. (1989), *Albrecht Dürer Melencolia I. Im Labyrinth der Deutung*, Fisher Taschenbuch Verlag.

Bornstein, D. (1978), *Distaves and Dames. Renaissance Treatises For and About Women*, New York: Scholar's Facsimiles and Reprints Delmar.

Braham, A. (1965), *Dürer*, London: Spring Books.

Calvesi, M. (1993), *La Malinconia di Albrecht Dürer*, Torino: Giulio Einaudi editore s.p.a..

Flanagan, S. (1991), *Hildegard von Bingen. A Visionary Life*, London: Routledge.

Klibansky, R., Panofsky, E. and Saxl, F. (1964), *Saturn and Melancholy. Studies in the Natural Philosophy, Religion and Art*, London: Nelson.

Levy, M. (1964), *Dürer*, London: Weidenfeld and Nicholson.

Maclean, I. (1980), *The Renaissance Notion of Woman. A study in the Fortunes of Scholasticism and Medical Science in European Intellectual Life*, Cambridge: University Press.

Musper, H.Th. (1952), *Albrecht Dürer. Der Gegenwartige Stand der Forschung*, Stuttgart: W. Kohlhammer Verlag.

Newman, B. (1987), *Sister of Wisdom. St. Hildegard's Theology of the Feminine*, Berkeley: University of California Press.

Panovsky, E. (1915), *Dürer's Kunsttheorie. Vornehmlich in Ihrem Verhältnis zur Kunsttheorie der Italiener*, Berlin: Georg Reimer.
Panovsky, E. (1943), *The Life and Art of Albrecht Dürer*, Princeton: Princeton University Press.
Schuster, P. (1991), *Melencolia I. Dürers Denkbilder*, Berlin: Gebr. Mann Verlag.
Wack, M. (1990), *Lovesickness in the Middle Ages. The Viaticum and its Commentaries*, Philadelphia: University of Pennsylvania Press.
Wölfflin, H. (1923), 'Zur Interpretation von Dürers Melancholie', in *Jahrbuch für Kunstwissenschaft*, Leipzig: Verlag Klinkhardt and Biermann.

Body-Guarded

The Social Aesthetics of Critical Care

Mary Morrison

The genesis of this project, BODY-GUARDED: The Social Aesthetics of Critical Care, *is founded in negotiating a terrain that incorporates the two disciplines in which I am trained; Fine Arts and Critical Care Nursing. On many occasions people have commented on the disparity of these worlds. I do not consider this so – both involve the human condition, and require observation and visualisation skills. Body-guarded refers to the protective needs of incapacitated patients, and social aesthetics to the use of art to examine social aspects of critical illness.*

In order to evaluate critical care artistically, I began what has become an ongoing journey. I have used a theoretical as well as visual approach, stimulated by an investigation of ideas that background the prevailing attitudes in hi-tech New Zealand hospital environments. To contemplate the use of unconscious patients as subject material, even with ethical approval and consent, seems inherently voyeuristic. Medicine uses the justification of medical science to conduct research into the pathology of the critically ill. I believe it is also valid to evaluate that human experience using fine art.

Critique

Critical care is an environment of extremes. To be there, one must qualify:

- patients – must be critically ill
- staff members – must have the personality and knowledge that copes with the pressures of incessant life-threatening events.

A patient presents to critical care, usually from the emergency department and/or the operating theatre. There, first-line interventions have transformed the person into patient and stripped them of their identity to the extent that they are named by their diagnosis.

In critical care, the significance of observation, visualisation and demonstration skills is fully utilised. The clinical primacy of 'seeing is believing' grants a diagnosis empirical verification.

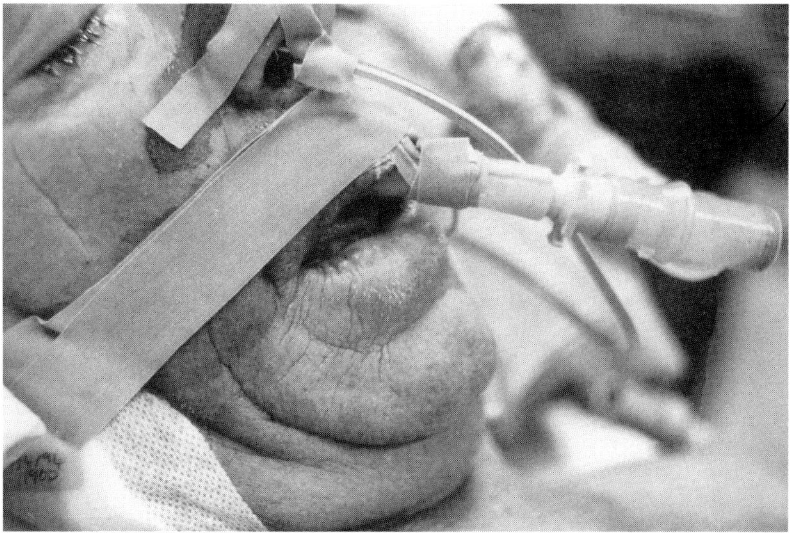

Mary Morrison, 1995

Tubed.

- The patient's body is horizontal, passive, partially naked and mute.
- It is tapped, felt, prodded, examined, listened to and compared.
- It is invaded for connection to requisite life-support and monitoring equipment.
- Its vital sign-language is translated by technology – ECG, EEG, ICP, CT and X-ray.

The seeing of a diagnostic eye is very much part of the negotiations in a critical care world, although in my opinion this visual literacy remains under-acknowledged within medical culture.

Textual theory supports the bedside assessment,
and what is seen creates the text.

The evolution of theoretical knowledge in medicine is founded in and derived from experience. Empirical data, collected and collated, can never really become intelligible unless grasped from an ideational frame of reference and viewed from a conceptual plane.

In our current hierarchy of knowledge, that which is text-based takes precedence over any other form. Barbara Maria Stafford, in her commentary on the exploration of Enlightenment art and medicine aims to

> ... expose how the visual arts and bodily-kinaesthetic intelligence in general, were damned to the bottom of the Cave of the Humanities. In today's text-based curricula, sensory and affective phenomena continue to be treated as second-rate simulations of second-class reflections.
>
> (Stafford, 1991: 2)

Cultivated intuition is born of the knowledge of experience.

Educated instinct becomes the legacy of a compilation of multiple examinations. It operates with the same kind of facility for both medical and artistic decisions.

- Artist – dissects an artwork
- Nurse – critiques patterns of breathing
- Intensivist – creates a percutaneous tracheostomy.

One learns to read the body in the same manner that one reads a painting or a sculpture, visualising the layers of meaning and modes of construction.

An-Aesthetic

The patient is unable to talk due to a reduced level of consciousness, illness or an endotracheal tube passed down through their vocal cords. Their responses to their immediate environment may be described as sensate or aesthetic perceptions.

Terry Eagleton states that 'Aesthetics is born as a discourse of the body' (Eagleton, 1990: 23). The origins of the word derive from the Ancient Greek *aisthitikos* which means *perceptive by feeling.*

> *Original site of aesthetics – is the reality of corporeal nature:*
> *taste, touch, hearing, seeing and smell are the senses, the effects of surface stimulation of the nervous system via the nose, ears, skin, eyes and mouth.*

The term aesthetics was first 'misapplied' to the criticism of artistic taste by Alexander Baumgarten in 1880. By the beginning of the twentieth century the term was used to describe the philosophy of taste and beauty. According to Susan Buck-Morss, the term had undergone a reversal in meaning so that:

> It was applied first and foremost to art – to cultural forms rather than sensible experience, to the imaginary rather than the empirical, to the illusionary rather than the real.
>
> (Buck-Morss, Susan, 1992)

Terry Eagleton describes the passage of change in the aesthetic concept from Baumgarten's critical materialist use to the sensibility of the aristocratic moral style of Burke and Shaftesbury. The sensuous high-style usage in the Germanic idealist myth of creative imagination from Kant to Habermas operates for Fredric Jameson as a 'sandbox' for vague things so they can be monitored and, if necessary, controlled.

> The aesthetic is in any case conceived as a kind of safety valve for the irrational impulses.
>
> (Jameson, 1990: 232)

By the late 1980s a new usage had evolved. 'Social Aesthetics' became associated with the concept of socially based art practice primarily intended to effect political change. Louis Althusser provided the theoretical pre-condition for conferring an activist status onto culture. He maintained cultural practices can possess critical powers as a direct consequence of the social struggle in class politics (Deitcher, 1990: 18).

The purpose of my project, *BODY-GUARDED: The Social Aesthetics of Critical Care* is social communication. Social aesthetics is used here to describe my use of art to examine the social issues of critical illness in contemporary New Zealand society.

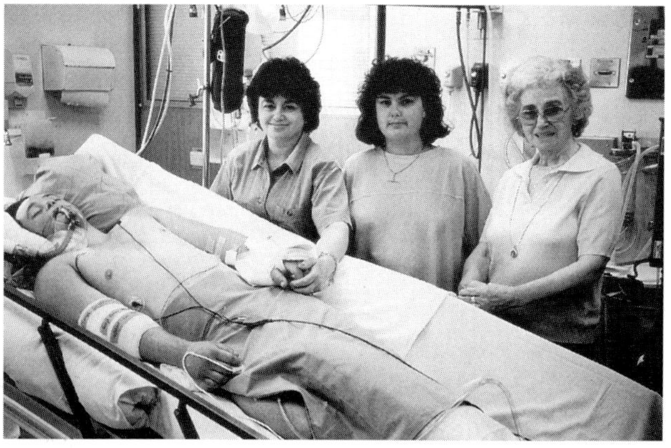

Mary Morrison, 1995

Family Affair.

> The great book of Man-the-Machine was written simultaneously on two registers: the anatomico-metaphysical register, of which Descartes wrote the first pages and which the physicians and philosophers continued and the techno-political register, which was constituted by a whole set of regulations and by empirical and calculated methods relating to the army, the school and the hospital, for controlling or correcting the operations of the body.
>
> (Foucault, Michel, *Discipline and Punishment*, vol. 1, p. 136)

Man-as-Machine

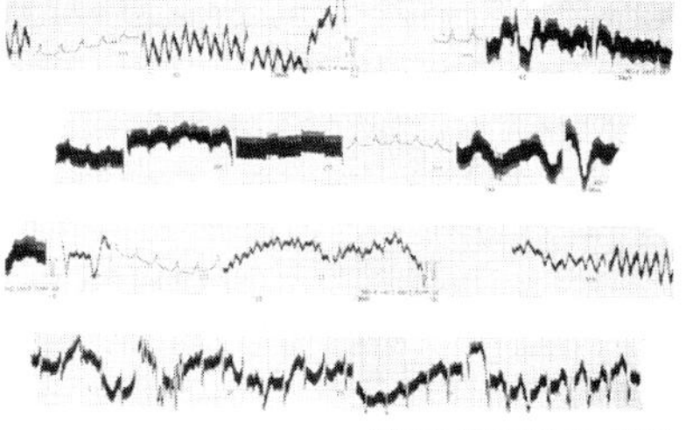

Mary Morrison, 1995

Rhythm Strip. (Cardiogenic Shock, ECG paper, 400 x 600mm.)

Foucault refers to the image of *man-the-machine* that was refined and developed in seventeenth century European thought. Previously, the soul and body had been considered united. The consequences of moral dramas, cosmological and social forces were reflected in the general health and wellbeing of the person.

By separating soul and body, the model of the body-as-machine translates to a medical body with ease. It can be understood as operating as a series of physical forces, such as electrical, chemical and kinetic. This idea in medicine was to influence the increasing focus on the specialised use of science, technology and machines to help repair the body-machine when it fails.[1]

René Descartes provided the philosophical and methodological basis for this shift in thinking. Dissection of animals became his primary method of gathering knowledge. Consequently, the dead body became a significant factor in his theories of embodiment.

He propounded, contrary to medieval thought, that the principle of vitality was to be found in the mechanistic processes of the body itself – not the soul.

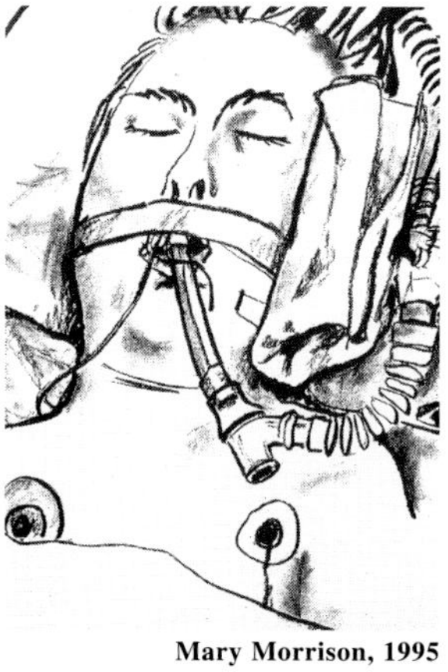

Mary Morrison, 1995

Collapse. (Charcoal on paper.)

> Death never comes to pass by reason of the soul, but only because some one principal part of the body decays ...
>
> (Descartes, 1970: 333)

In *Description of the Human Body and all its functions, 1648*, Descartes provides an analysis of voluntary movement, using the explanation that the body obeys the soul only on the condition that the soul is primed mechanically to do so.

> The soul cannot produce movement without the appropriate disposition of the body organs which are required for making body movement. On the contrary, when all the bodily organs which are required for some movement, ... the body has no need of the soul in order to produce the movement.
>
> (Descartes, 1967: 315)

Descartes does not use a political analogy of a chain of commands for the soul to move the body. Rather it is a mechanistic view of the organs acting as a series of gear wheels.

Georges Canguilhem takes this notion further recently, by suggesting that machines can be considered as organs of the human race (Canguilhem, 1992: 55).

> *A tool or machine is an organ and organs are tools and machines.*

This implies a sense of rationalisation and standardisation. The machine functions within a set of rational economic rules, measurable standards and interchangeable parts. Using this logic, a failed body organ may be replaced by its mechanistic equivalent and be resited outside the body, such as an artificial kidney dialysis machine or ventilator breathing machine, on a long or short-term basis.

To understand an object in mechanistic terms is to break it down into its separate elements and analyse their interaction. By considering the machine as a set of data, with the application of knowledge, it is possible to resolve problems by starting with an idea and working logically and systematically through that data.

Once the elements and forces in the process are understood, the power to alter them and duplicate them is available. When nature is conceived of as a machine the same principles apply and previous constraints or prohibitions may be removed (Leder, 1992: 21). There is therefore a crucial shift from the passive contemplation of the whole of nature, including other life forms, to one of active manipulation.

Medicine and the Body of the Corpse

> *The dead human body can easily be equated with the lifeless machine.*

Modern medicine is profoundly Cartesian in spirit. It has been shaped by the body of the corpse that stimulated Descartes' explorations, rather than the living body. This focus has had the effect of shaping disease categories, diagnostic techniques and modes of treatment.

In the eighteenth century, classification of diseases shifted from a basis of symptoms experienced by a living patient to a basis of organic lesions found on the corpse. The human body became the object or space in which the disease was located. With the body subtracted, the medical gaze was directed toward the disease itself and the signs and symptoms which differentially identify it.

In classificatory medicine, death constituted the end of the disease and of life. In contrast, in anatomo-clinical medicine, death provided an analytic point in which to examine life and disease, organic and pathologic consequences (Foucault, 1973: 10).

> The question 'what is the matter with you?' with which the eighteenth-century dialogue between doctor and patient began ... was replaced by that other question, 'Where does it hurt?'
>
> (Foucault, 1973: xviii)

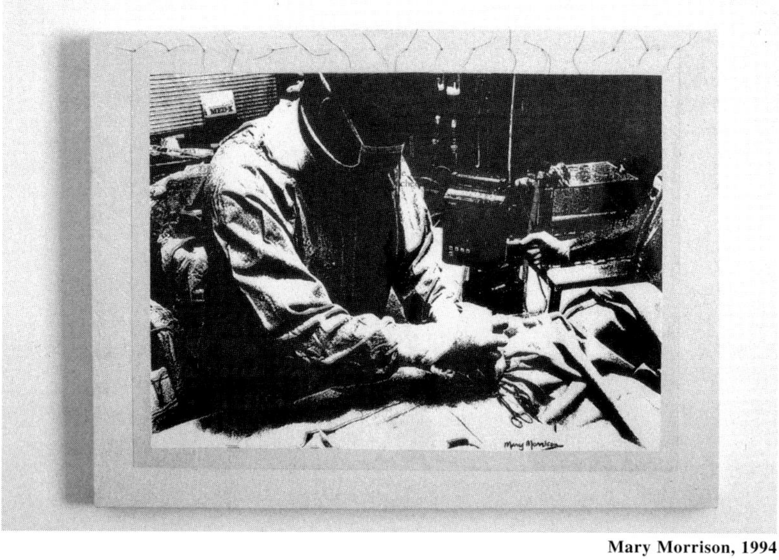

Mary Morrison, 1994

Cut Throat. (Photosilkscreen on canvas, clear vinyl and suture material, 650 x 850 mm.)

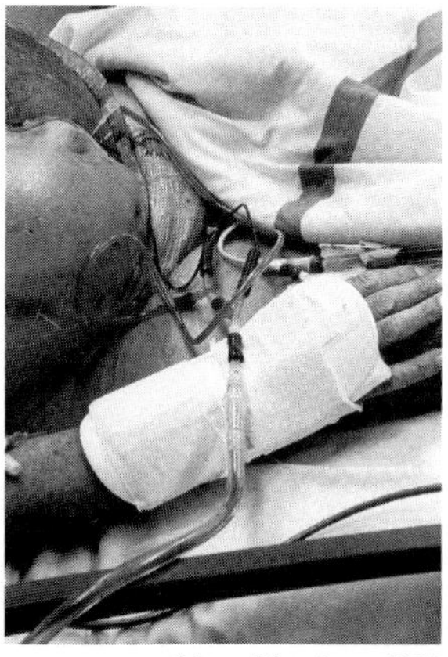

Mary Morrison, 1995

Abdominal Split.

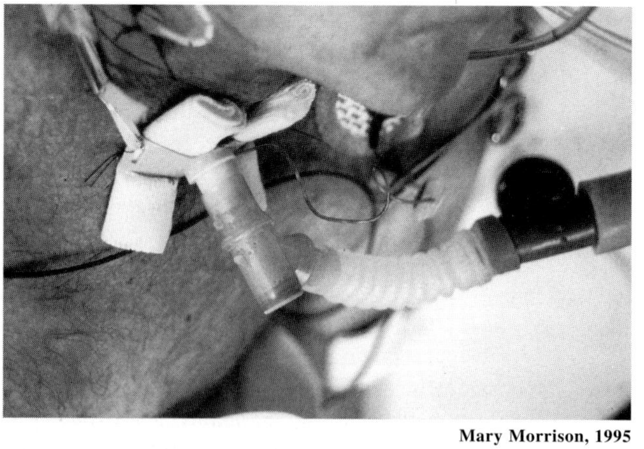

Mary Morrison, 1995

Blow.

The introduction of the individual into medical discourse in anatomo-clinical medicine as both subject and object of knowledge provoked the conceptualisation of distinctions between normality and pathology.

The idea of normality had existed since Vesalius' sixteenth-century anatomy lessons. It was customary to ensure that the cadaver used for public dissections was as normal as possible (a middle or general form). Other bodies may then be assessed comparatively, pointing out the deviations from that central norm. Normality could then be defined as the absence of pathological symptoms in organs.

Abnormality started to be related to degeneracy: the single body predicated a population and thus the programme of medicalisation began. This transformation of general social problems into technico-medical concerns confirms the potential for the use of medicine as a means of social control.

The Socio-Political Body

The body-as-machine, Michel Foucault suggests, is not only a metaphysical position traced by Descartes, but provides a logic facilitating the political control of human beings. The population increases in the eighteenth century provoked the necessity to place the body of the individual and the body of the population under the *microscope of behaviour* for its regulation, co-ordination and productive management (Foucault, 1980: 171).

Once people are conceived of as machines,

- worker machines
- sex machines
- soldier machines ...

they can be manipulated and trained for precise social functions.

In Foucauldian thought, power unfolds within society through large-scale events (in the political arena) as well as through complicated *micro-physics*. Power is distributed through the home, hospital, workplace, school and the barracks and operates on humans. The norms of health and illness advanced in medicine and the settings where treatments take place – hospitals, clinics and so forth, serve an important role in this economy.

Foucault's interest is in the knowledge relations across institutional boundaries, that is the effects of medical practice on social life rather than in the knowledge capacity of medical science.

Medical practice is double-sided; offering qualified help and considering disease as a political and economic problem.

In the context of medical practice, Foucault maintains that the patient's body may be exploited or traumatised due to its place at the bottom of the clinical hierarchies, but the patient is not as a result, any more or less controlled by power relations than are the hospital staff.

When the patient enters the medical domain for treatment and cure, their behaviour must proceed in accordance with the clinical protocols of that disease and hospital in order to return to health. The staff, architectural design, equipment, schedules, costing regimes and so forth, are all caught up within the same web of controls.

The power relations are not so much between the doctor and the patient, but the multiplicities of practices and features within the socio-medical discourse (Spitzack, 1992: 64). The authority of the physician is both underscored and undermined by the demand to identify and treat illness.

Similarly, in the instance of incompetency, a healthy family member can easily become a proxy and take on the identity of the ill relative, in order to refuse or accept treatment on their behalf.

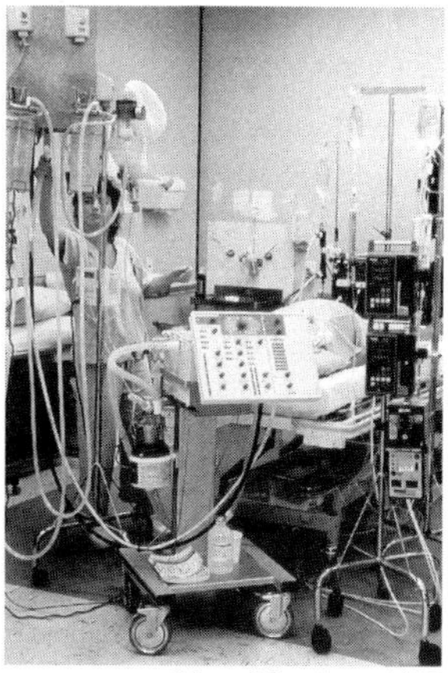

Mary Morrison, 1995

Bedspace.

Fabric of the Body

When a new patient is received into critical care, their personality and clothing have already been stripped away. Initial assessment and interventions by the hospital emergency department and/or the operating theatre ensure that the patient's failing body is presented as only that. The failing body. And to that purpose the critical care staff intervene.

For the critically ill, the body surface is their clothing. The need to visualise beneath this surface is imperative and urgent and to do so various procedures are rapidly instigated. The mechanistic body is opened up radiologically, electromagnetically and surgically in true empirical spirit.

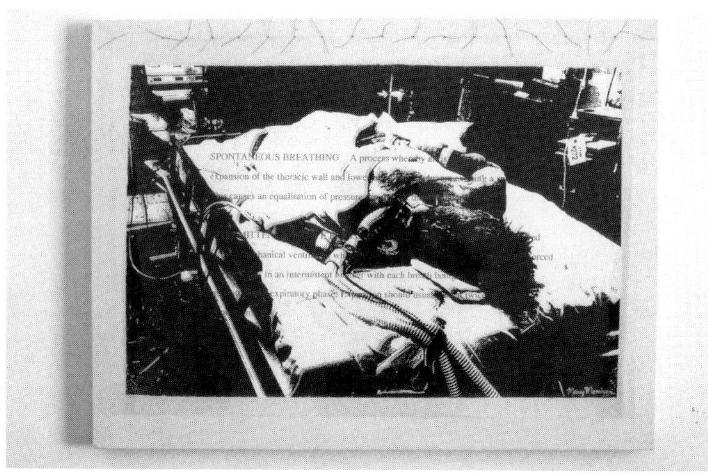

Mary Morrison, 1994

Downhill. (Photosilkscreen on canvas, clear vinyl and suture material, 130 x 1700 mm.)

To reveal knowledge, classic nudity is no longer sufficient.

The body surface can only partially operate as a garment. Skin is penetrated both invasively and non-invasively and the body is laid open to unveil the secrets of the failing body-machine.

George Bataille, in his analysis of eroticism and its relationship to death, maintains that,

> The urge towards love, pushed to its limit is the urge towards death.
>
> (Bataille, 1957: 42)

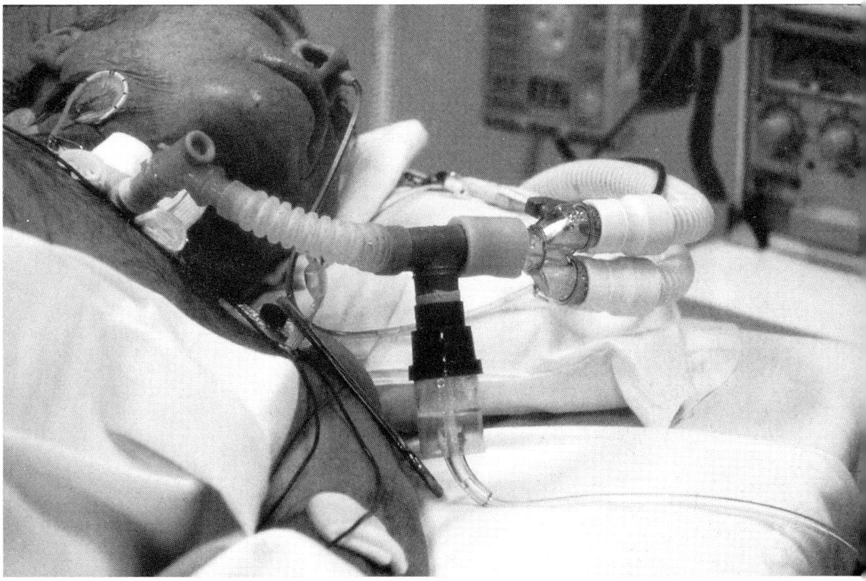

Mary Morrison, 1995

Lucky.

To wound and expose is to lose oneself in an abyss and multiply desire with the enactment of transgressions. Erotic power surrounding the violation of social taboos is stimulated by desire of infringement. For Bataille, death and sexuality in abundance and excess, exchange their energies and excite each other.

> I can tell myself that repugnance and horror are the mainsprings of my desire and such a desire is only aroused as long as its object causes a chasm no less deep than death to yawn within me and that this desire originates in its opposite, horror.
>
> (Bataille, 1957: 59)

Whilst an unconscious or dying patient may conform to Bataille's categories, the patient's abject situation challenges his theoretical position. At some point dire illness sets limits which preclude desire and eroticism. The extremes of Bataille's metaphysical notions are pulled back by the harsher parameters of reality.

The American artist Andres Serrano has produced a series of large-scale colour photographs entitled *The Morgue (Cause of Death), 1992*. The documentation of death is sublime in these larger than life Cibachromes (1270 × 1520 mm) and the viewer is able to art-historisize the images. Presented in close-up detail, the grandeur of a surgical incision in *The Morgue (Rat Poison Suicide II), 1992* becomes stigmata.

To see death in public is startling. The experience of dying in the domestic realm has gradually been removed by bourgeois society, to the confines of the appropriate social and hygienic sites – the hospital, the morgue and the funeral parlour.

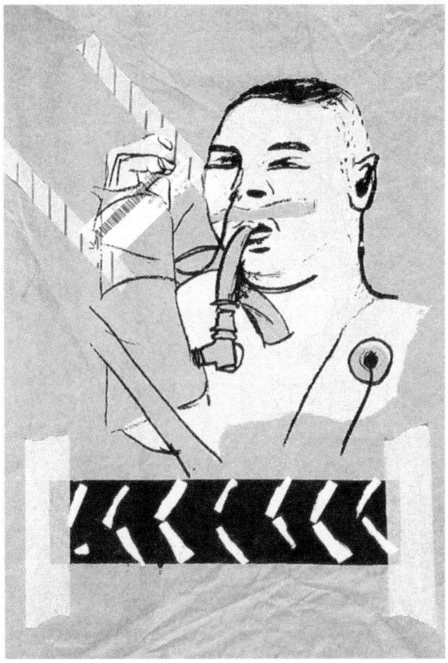

Mary Morrison, 1994

Crash – Portraits-in-Absence detail.

Mary Morrison, 1994

Bugged – Portraits-in-Absence detail.

Silkscreen and stitching on green sterile hospital paper, 720 x 3950 mm.

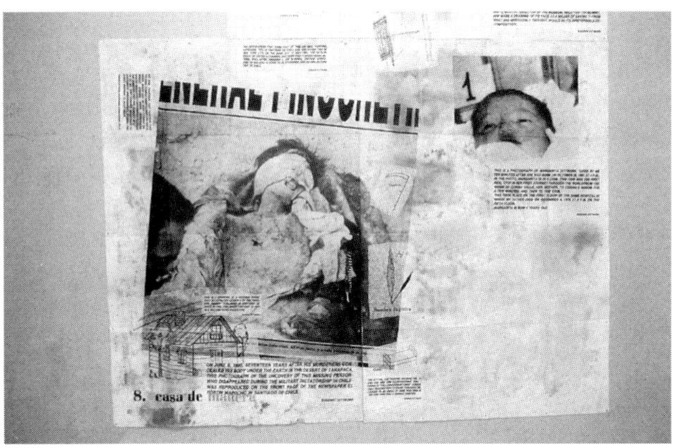

Wellington City Art Gallery

Airmail painting No. 90.
1994 Eugenio Dittborn paint, feather, stitching and photo-silkscreen on three sections of non-woven fabric, 3500 x 2800 mm.

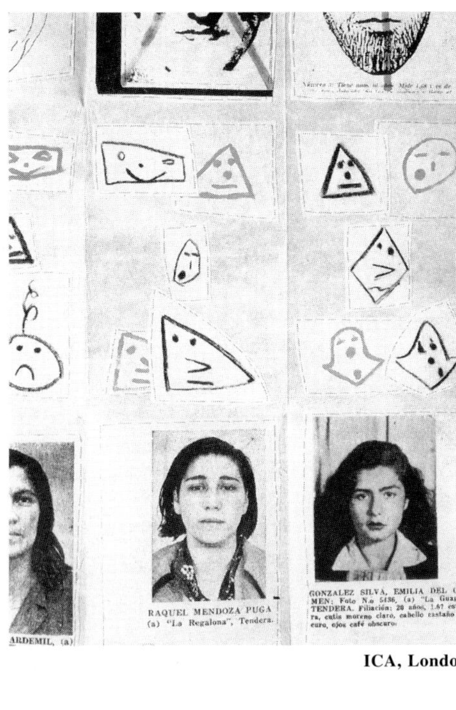

ICA, London

Airmail painting No. 78.
The 7th History of the Human Face (The Scenery of the Sky) *1990 (fragment). Eugenio Dittborn paint, stitching, charcoal and photosilkscreen on three sections of non-woven fabric 2150 x 2800 mm.*

Serrano's desire to achieve a final heroic monumentalisation in these photographic scenes of death is thwarted by the overt sense of eroticism and objectification. Horizontal poses, draped fabric, cropped compositions, wet lips, uncovered skin and glossy surface all conspire against his intentions and pose a troubling allure.

Bataille states:

> The generative power of corruption is a naïve belief responding to the mingled horror and fascination aroused in us by decay.
> (Bataille, 1957: 56)

Death is death, an act of severance. The invitation extended by Serrano to the viewer in these images is ongoing and problematic. Even the decaying blood vessels in *The Morgue (Death by Drowning), 1992* have become decoration, a pattern on the skin, just like the tattoo already present. One knows nothing of these people, we are no closer to their lives, or to those who have survived them. You cannot deny the attraction of their beauty, but you can deny their death.

The Dead Christ and Critical Care

The bedspace which contains the patient and his or her life support equipment is a volume of space. Its dimensions relate directly to the proportions of the prostrate body horizontal in the bed. The patient, isolated by illness lies there, simply covered with a white sheet.

This act of composition is an act of isolation, similar in manner to Hans Holbein, the Younger's *The Body of Christ in the Tomb (1522)*. There, Christ's horizontal body fills out the canvas to the edges, a larger than life-size corpse. His head is rotated slightly towards the viewer and his hand beckons – both are tinged green by mortality. The stigmata of the Cruxificion and the puncture wound left by a spear, mark his emaciated body. It rests stiffly on a hard slab.

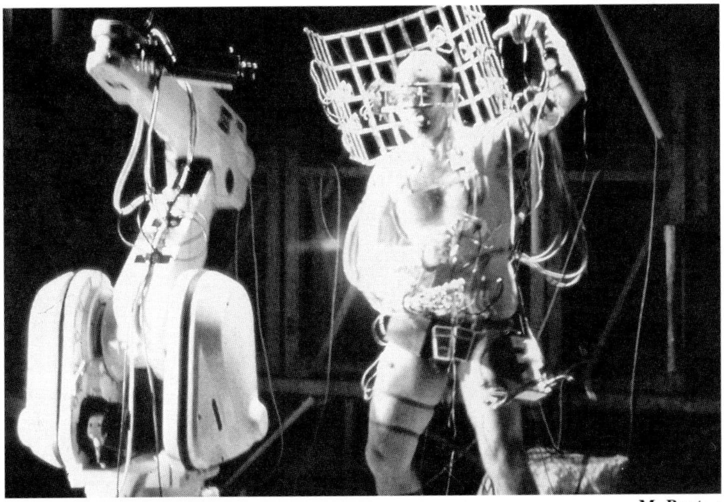

M. Burton

Scanning Robot/Involuntary Arm, Stelarc Edge Biennale, London.

It is the face of a man recently dead. A look of suffering is still discernible and a certain slackness in his expression confirms that his dying is over. The portrayal of death is commonplace, real and approachable. This passive state provides a link to the critical care patient. Both are in limbo temporarily before some kind of transition occurs. Christ was resurrected; and critical care allows one of only two possibilities: to live or to die.

For the relatives entering critical care, there is the initial shock of walking into an environment that makes no distinction between night and day. They hear the sounds of high-pitched alarms and the regular whoosh of the ventilator breaths. Through the vine-like array of tubing and the blockade of high-tech electronic equipment lies their distorted, comatose family member. As they approach, an initial reaction of denial is in conflict with a protective desire towards the patient.

The nurses' role is to demystify, mediate and incorporate the relatives into this temporary realm on behalf of the patient. They gently guide the relatives through to some kind of understanding, answering questions and listening to any stories about the patient. The stories and perhaps a photographic snapshot add up to an abbreviated portrait of the person who up to that point for the staff has been very much a reflection of their diagnosis.

Whilst the procedures are being carried out on the patient, the critical care staff will continue to inform the patient of what is happening despite their seemingly reduced level of consciousness. This operates as a courtesy to the incumbent patient and brings some kind of humanity into the bedspace.

To deal with the body of the patient as a carcass is psychologically implausible. The constant pressures, the initiation of (often brutal) procedures and the responsibilities of looking after the critically ill take their toll. The staff need to feel that the job they are performing is worthwhile and that there is a person underneath the illness whom they are helping to return to health, or alternatively facilitating a good quality and compassionate death.

Simon Hunter

Event for the Stretched Skin/Third Hand (The Last Suspension). 1988 Stelarc Ofuna Monorail Station, Yokohama.

Somehow in critical care everyone, well or ill, gains a tactile mortality. In an attempt to evade emotional stress, a black hospital humour has evolved, a macabre pleasurable irony based on medicalised and grotesque predicaments.

In Holbein's *Danse Macabre (1524)*, a series of woodcuts commissioned for European circulation, one sees the same humour. Death, represented by a laughing skeleton, indiscriminately moves through all levels of society to embrace its target. No one, not the bishop, the idiot, the rich or the noble woman escapes. Sarcasm and mockery are defiant weapons because to laugh in the face of death is to deflate the power of death to overwhelm. (And then you cry.)

Tony Fomison painted two studies of Holbein's *The Dead Christ* in the early 1970s *(Untitled) Study of Holbein's 'Dead Christ'* and *Second Study of Hans Holbein the Younger's 'The Dead Christ'*. Here Fomison presents to the viewer the anguish of Christ during his dying process, rather than the commonplace physicality of Christ's life-end in Holbein's version.

The inverted psychological tension in *Second Study* relates closely to those patients who are intubated and ventilated, but conscious. Too ill to really process what is happening to them and unable to communicate, theirs is an existential struggle, such as we see in Fomison's *Christ*.

The figure in the Holbein had been modelled from a corpse fished out of the Rhine River, a document of decay. Perhaps Fomison lacked the opportunity to paint from death, which precluded the sense of finality and limits of death in his *Christ*. He had worked from a life-model as well as Holbein's painting (Wedde, 1994: 261). In reflection, Fomison has the viewer contemplating the reality of their own mortality.

The break of the bond of life between Christ and His Father is,

> ... a mythical representation of a fundamental and psychically necessary discontinuity,

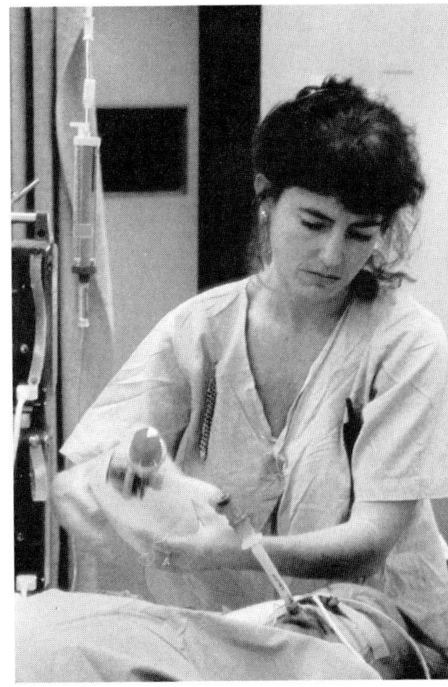

Mary Morrison, 1994

Breather.

according to Kristeva (Kristeva, 1989: 261). In psychoanalysis a series of separations – real, imaginary or symbolic – structure our individuation: birth, weaning, separation, frustration, castration.

The same maternal language and processes occur as the critical care patients regain their autonomy. When *weaning* from the ventilator breathing machine, the ventilator work-load is reduced and conversely patients are required to make more respiratory effort. The inotropic drugs administered continuously to support blood pressure, are gradually *weaned* off as the cardio-

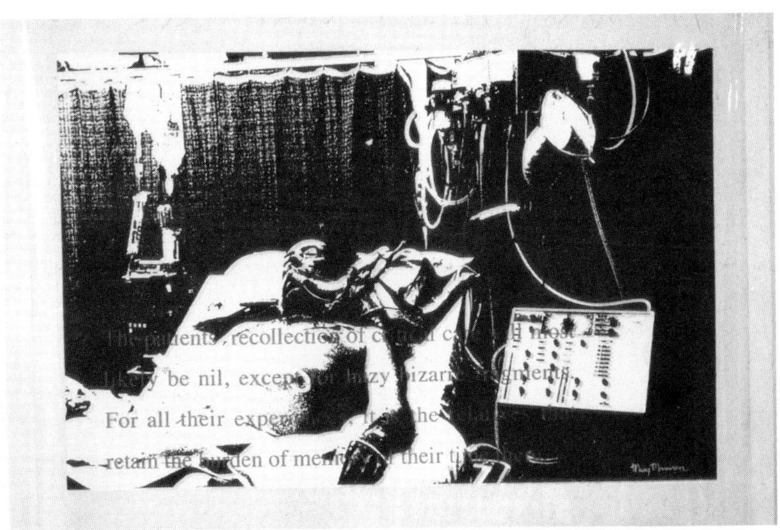

Mary Morrison, 1994

Time Lapse. (Photosilkscreen on canvas, clear vinyl and suture material, 1300 x 1700 mm.)

vascular system stabilises. Whilst the patients become more alert, the more frustrated they become and the more frustrating to look after.

These are all positive signs. The support and monitoring equipment are progressively removed from the patients who are now ready for transfer to the wards. Their recollection of critical care will most likely be nil, except for bizarre hazy fragments.

Dreams and reality superimpose to produce an attenuated memory. To one patient, the roar of a suction catheter being passed, and subsequent cough, became the roar of a passing Bristol freighter which dropped a load of pollen onto him causing a coughing fit. High body temperatures correlated with another's sensation of being left for hours and days on the edge of a desert. Just when he could stand it no longer, he recalled being pulled into the shade. Another thought that mouth-cleaning swabs were shrimps and kept trying to eat them.

A sense of any chronological sequence, of one day following on from another, may be lost. The markers of time are procedural only – hospital interventions that interrupt their isolated states. For all their experiences it is the relatives that retain the burden of memory of their family member's time in critical care.

I would like my *Portraits-in-Absence* to function as mementos of lost time for the critically ill. They are sketch-portraits of intubated patients, silk-screened onto green sterile hospital paper and stitched together. Their level of unrecognisibility as a likeness of the individual is such that their primary meaning is metaphorical.

These images relate in one sense to the memorial aspects of Eugenio Dittborn's portraits of the *disappeared* in Chile. His light-weight, large-scale silk-screened montages are folded up into their airmail packages and posted out of Chile. The journey of the *Airmail Paintings* functions as suspended time, like the journey experienced by patients in critical care.

Operating through the repressive Pinochet regime, Dittborn used his art to critique the Chilean political system, the power of the state and the surveillance role of the camera. Dittborn describes,

> ... the political aspect of my work was to be found in the folds of the Airmail Paintings (like a poisonous powder hidden there).
>
> (Dittborn, 1993: 2)

The military regime declared that pre-1973 nothing existed – the collective memory was censored. By disinterring old newspaper articles and police photographs of petty criminals (mostly dispossessed and impoverished farmers) to include in his art work, Dittborn disrupts, displaces and remembers. He makes his images travel through time to recombine *and unearth acts of memory*.

> *Each image breaks the silence of forgetting in response to other images which pose it pointed questions.*

The Obsolete Body

Stelarc is an Australian performance artist, whose investigations into issues of information and technology in relation to the body have led to his theory of the obsolete body. In a post-industrial society he defines the body as overloaded and trapped, not by gravitational pressure, but in a technological crush of specialist information.

In its present form, Stelarc maintains, the body is obsolete, but still capable of evolving with Larmarkian speed to form a new human hybrid when stimulated by the products of miniaturised medical technology attached and/or implanted in the body. Rapid evolution becomes a conscious process in which artistic intervention is a necessity and artists may become the evolutionary guides (Paffrath, with Stelarc, 1984: 24).

Stelarc's performances are intended to elaborate on these theories. In one type, he attaches an electro-mechanical third arm to his right shoulder. This is operated by electrodes that translate the movements of his abdominal muscles into an electric current which stimulates the prosthetic device. One of his original arms, in response to an electrical charge, moves with wild random jerks. Such is his skill at operating all three arms together, that he can write a continuous, albeit simple, sentence.

He has choreographed an elaborate computerised performance between himself – with third arm attached and flashing laser eyes and a heavy industrial robot. This exhibition has all the pleasures and dangers of a large mammalian mating dance.

Another performance involves a series of suspensions in which Stelarc *supports the body in space*. He uses approximately 18 hooks that are inserted through the skin, fat and muscle layers of his torso, calves and upper arms. The cord, attached to the hooks, is connected via a pulley system to a wooden structure. He may place electrodes on his body to engage the audience in a biofeedback system of amplified sounds of muscle and heart rate. These suspensions, conducted over a period of 20 years and governed by pain-tolerance, may last anywhere from 60 seconds to 25 minutes.

The association with the tradition of suffering is reduced by the masochistic edge to the performance. Stelarc denies he feels any pleasure, but he certainly feels pain. The limitations of his body seem to be related to his ability (or inability) to endure the pain.

He has refrained from using any mind-control techniques or anaesthesia, mimicking the conditions that held back the development of surgery in the pre-anaesthetic era, prior to the mid-nineteenth century. The advent of an induced loss of consciousness allowed surgeons to operate without emotional involvement and patients no experience or memory of the pain associated with the operation itself (Jordanova, 1989: 136).

Stelarc is always master of his own destiny, directing the performance to the end. The critical care patients are subject to much greater forces of fate, illness and the hospital. Their suspensions are of time, ultimately longer and more painful, which terminate by cheating or succumbing to death – a matter of survival, not evolution.

The Plastic Body

Orlan is a French performance artist who utilises the application of surgery for the purposes of culture not pathology. Since 1990, in her bid *to break open the bland discourse of appearance*, Orlan has staged a series of 12 performance-operations in which her face and body are remodelled to comply with *the reincarnation of Sainte-Orlan.*

The blueprint used by the surgeons is a compilation, or *cross-breeding,* of features from women whom Orlan has selected from art historical sources,

> ... not for their beauty but for their characteristics and their stories.
> (Bond and Orlan, 1992: 184)

Orlan, so far, has

- the forehead of Da Vinci's *Mona Lisa*
- the chin of Botticelli's *Venus*
- the eyes of Gerome's *Psyche*
- the mouth of Boucher's *Europa*
- a nose from the School of Fontainbleau.

(Wark, 1994: 63)

To finance her transformations she markets videos and photographs of the performances, along with reliquaries containing excised flesh and bodily fluids.

Her performances are staged within an operating theatre which has been draped with decorations and sterile fabric theatre-guards. The surgical team wears a mixture of theatre scrubs and Renaissance-style costumes, designed by Orlan's stylist. Surgery is completed under local and/or epidural anaesthesia. Orlan remains conscious to direct the video camera or read out loud from theoretical texts whilst on the operating table.

The surgery becomes a surgical farce in the face of Orlan's abandonment of pathological need to cultural necessity. Laura Cottingham refers to her commodification,

> or perhaps prostitution is the more apt metaphor: First Orlan gives her body over to authority, then she sells it.
> (Cottingham, 1994: 60)

Orlan's project, despite the publicity, videos and the theatrical posing, does not open out the private world of medicine. Medicine is aligned to the realms of confidentiality, secrecy and privacy and this costumed display acts as a disguise, rather than a revelation. The aura of privilege and control over knowledge is reinforced as she submits her body to the surgeons who are able to transform it into its 'idealised' state. Their professional distance from the public is maintained.

Her new beauty is skin deep, this time the hand of the male artist has been replaced by the doctor's scalpel. The confusion between image and self, the *bland discourse of appearance* that Orlan wishes to challenge, holds all the appeal of an in-depth 'before' and 'after' magazine makeover. By contemporary standards her project can never be complete, not least due to the inexorable ageing of her body. The cultural conventions remain intact from the Renaissance to the Post-industrial society, with the relationship of the viewers' gaze to the objectified female body undisturbed.

The collusion between the media and the gallery regarding the publicity that surrounds her performance-operations opens another debate. Without the pre-arranged publicity, would the interest in this reincarnation fall away? Art becomes a means for self-transformation of the self, not visa versa. Does Orlan show up cultural attitudes, or is she shown up by them? Ultimately one must question her motivation, fascinating as the performances are.

In Conclusion

The model of body-as-machine has given rise to therapeutic triumphs, as well as to limitations and distortions in medical practice. That modern medicine often neglects the importance of psychological factors and beliefs in the etiology and treatment of disease, is almost a cliché.

Cartesian medicine can be transformed and extended by the incorporation of other ways of evaluating human experience. This broader framework would provide for the emergence of a new paradigm. As artists are evaluating the empirical worlds of science and medicine, they too have much to examine and contribute.

My predominant concern has been with the dynamics of two apparently very different sites. My aim has been to open out to view the closed and privileged environment of critical care, by revealing parallels in terms of practices and dynamics with fine arts.

Specifically:

- Visual literacy as a body of knowledge which is operated by intuitive and trained aesthetic judgements.
- The utilisation of conceptual and creative ways of thinking. Patients and art work are both routinely and formally critiqued in critical care and art practice.

To meet an ex-patient and have their personality reassert itself, evident in their facial expressions, the unexpected sound of their voice and lively eyes, often comes as a surprise. As a guardian of their lost time, I realise I never actually knew them.

> Dear Mary
> We are getting married at ... Henderson at 2 o'clock on Sunday
> 4th Sept. John and I would like it if you could make it.
> Cheers, Pauline
> Amen to the above, John.

Epilogue

The Politics of Outcome

Decision-making in New Zealand hospitals is increasingly driven by political and economic forces, rather than motivated by medical knowledge, compassion and need. There is a curious inability on the part of health administrators and politicians to acknowledge that the nature of health care runs counter to the present capitalist notions of profit and loss. The critical care patient requires the largest short-term investment, yet is surely an investor's nightmare.

One need only to look over the Pacific Ocean to where the American commercialisation of medical knowledge has created a health system in crisis. The relative costs of treatment in a United States critical care unit are three times that of New Zealand's.

> The extensive privatisation of scientific medicine, the growing importance of commercial chains of hospitals, the super-profits of the pharmaceutical industry, the dominance of managerial norms of performance in health-care delivery and the general commodification of health have major implications for the professional standing of doctors who may be increasingly subordinated to the position of employee of large health organisations.
>
> (Starr, 1982: 82)

The American attitude towards a patient is inherently different from the New Zealand one. Insurance payments, legal and profit factors all have an impact on the type of care delivered.

American citizens believe that if they want to, each and every individual is entitled to admission and treatment in a critical care unit, regardless of the nature of their illness. In New Zealand, patients are accepted into critical care by an Intensivist – when there is a reasonable chance of reversing the illness and restoring them to health, not when their situation is clearly futile. If their situation is uncertain, limitations for treatment may be placed and if it becomes hopeless, humane care is provided. This may involve the withdrawal of intensive therapies and allowing nature to take its course (Streat and Crippen, 1995).

Note

1. Meanwhile early thermodynamicists were struggling with the non-idealised behaviour of machines. Carnot (1827) produced a statement of the second law before most of the necessary base constructs had even been identified (internal energy, entropy and so forth), a remarkable achievement. Subsequent developments which undermine the man-as-machine paradigm include statistical mechanics (states defined in terms of the most likely – turn of the century) and non-linear mathematics (chaos theory – 1970s) which highlight the non-deterministic nature of most real systems. Only economists still unreservedly carry the torch high for the machine ideal these days, as to admit dissipative processes as inherent, or non-predicability as common would mean almost completely rewriting their field, with attendant marked loss of status. Mainstream economics fails first-law as well as second-law constraints and is clearly our most impressive attempt to create mechanistic perpetual motion. The rewrite is happening in highly marginalised circles: non-linear economics, evolutionary economics and ecological economics and notions of real risk in society.
   (From a conversation with Robbie Morrison, a Wellington environmentalist and energy activist.)

References

Bataille, G. (1957), *Eroticism, Death and Sensuality*, San Francisco: City Lights.

Bond, A. and Orlan (1992), 'Orlan – France', in *The Boundary Rider – 9th Biennale of Sydney*, Sydney: Art Gallery of New South Wales.

Buck-Morss, S. (1992), 'Aesthetics and Anaesthetics: Walter Benjamin's Artwork Essay Reconsidered', in *October*, vol. 1, no. 62, pp. 3–41.

Canguilhem, G. (1992), 'Machine and Organism', in Crary, J. and Sanford K. (eds), *Incorporations*, New York: Zone, p. 55.

Cottingham, L. (1994), 'Orlan', in *Frieze*, vol. 1, no. 14, p. 60.

Deitcher, D. (1990), 'Social Aesthetics', in *Democracy/A project by Group Material*, Wallis, Brian (ed.), Seattle: Bay Press.

Descartes, R., *Philosophical Letters*, Kenny, A. (trans and ed.) (1970), Minneapolis: University of Minnesota Press.

Descartes, R., 'Description of the Human Body and all its Functions', in Haldane, E. and GRT R. (trans) (1967), *Philosophical Writings of Descartes*, New York: Cambridge University Press, vol. 1, p. 315.

Dittborn, E. (1993), *Mapa: The Airmail Paintings of Eugenio Dittborn 1984–1992*, London: ICA.

Eagleton, T. (1990), *Ideology of the Aesthetic*, London: Blackwell.

Foucault, M. (1973), *The Birth of the Clinic*, Sheridan Smith (trans), London: Tavistock.

Foucault, M. 'The Politics of Health in the Eighteenth Century', in Gordon, B. (ed.) (1980), *Power/Knowledge, Selected Interviews and Other Writings 1972–1977*, Brighton: Harvester Press.

Jameson, F. (1990), *Late Marxism: Adorno, or, the Persistence of the Dialectic*, New York: Verso.

Jordanova, L. (1989), *Sexual Visions, Images of Gender in Science and Medicine Between the Eighteenth and Twentieth Centuries*, England: Harvester Wheatsheaf.

Kristeva, J. (1989), 'Holbein's Dead Christ', in Feher, M. (ed.), *Fragments for a History of the Human Body, Part One*, New York: Zone, p. 261.

Leder, D. (1992), 'A Tale of Two Bodies', in Leder, D. (ed.), *The Body in Medical Thought and Practice*, Netherlands: Kluwer Academic Publishers, p. 21.

Paffrath, J. with Stelarc (1984), *Obsolete Body/Suspensions/Stelarc*, California: J.P. Publications.

Spitzack, C. (1992), 'Foucault's Political Body in Medical Praxis', in Leder, D. (ed.), *The Body in Medical Thought and Practice*, Netherlands: Kluwer Academic Publishers.

Stafford, B.M. (1991), *Body Criticism, Imaging the Unseen in Enlightenment Art and Medicine*, Massachusetts: MIT.

Starr, P. (1982), *The Social Transformation of American Medicine*, New York: Basic Books.

Streat, S. and Crippen, D. (1995), 'Witholding and Withdrawing Life Support Systems in the Intensive Care Unit: U.S. and N.Z. Perspectives', Auckland: Auckland Hospital, Department of Critical Care Medicine, unpublished lecture.

Wark, McKenzie (1994), 'Post Human? All Too Human', in *World Art*, vol. 1, no. 1, pp. 58–65.

Wedde, I. (1994), *Fomison, What Shall We Tell Them?,* Wellington: Wellington City Art Gallery, p. 261.

CONTRIBUTORS

Heleen Buijs is an Art Historian and Lawyer. In 1996 she was Honorary Lecturer at the Department of Women's and Gender Studies at the University of Waikato (Hamilton, New Zealand). Her research interests are concerned with the mutual influences between Renaissance and Northern European artists in the early sixteenth century on the one hand, and modern art on the other. She is currently working on her doctoral thesis on the position of photography within the Surrealist movement.

Jenny Carryer is a registered nurse working currently as a Lecturer in the Department of Nursing and Midwifery at Massey University (Palmerston North, New Zealand). She is nearing completion of her doctoral thesis and teaches and researches in the area of women's health. She is keen to see the development of areas of nursing practice informed by feminist analysis and will continue to research and write in those areas.

Dr Marion E.P. de Ras is Professor of Women's and Gender Studies at the University of Waikato (Hamilton, New Zealand). One of her research interests concerns the historical and cultural construction and genesis of girlhood, girls' movements, girls' health and so-called girls' diseases. She has published several articles and books in this area. In 1993 she organised *Alice in Wonderland*, the first international conference on girls and girlhood, in Amsterdam. She has organised several conferences on the topic of gender, the body and medicalisation.

Susan Duncan is a registered nurse currently working as a Staff Development Co-ordinator for a Rest Home and Hospital focusing on the care of the elderly in Hamilton (New Zealand). Susan's doctoral work focuses on the medicalisation of sexual difference. Her current research interests include the ageing process(es) and death and dying.

Dr Victoria Grace is senior lecturer in Feminist Studies at the University of Canterbury (Christchurch, New Zealand). A recently completed research project on the problems women encounter in consulting doctors for chronic pelvic pain has led to an interest in a current project studying medical and biomedical discourses on endometriosis. She is also writing a feminist reading of the work of Jean Baudrillard.

Dr Kathryn Matthews Lovering is currently an independent scholar based in London and Cornwall as well as Senior Research Fellow at the Department of Social Policy and Social Work, Faculty of Human Sciences, University of Plymouth (U.K.). She has published on the topic of adolescents' talk about menstruation. She is currently doing research on adolescence, menstruation and the social sciences as well as a longitudinal study with girls on their experiences of menstruation.

Dr Judith Macdonald lectures in Social Anthropology at the University of Waikato (Hamilton, New Zealand). She carried out her fieldwork with the Tikopia, a Polynesian outlier community in the Solomon Islands in 1979 and 1980. Her research interests are Polynesian ethnography and women's health, both in New Zealand and in cross-cultural contexts.

Dr Emily Martin is Professor of Anthropology at Princeton University (U.S.A). Beginning with *The Woman in the Body: a Cultural Analysis of Reproduction* (Beacon Press, 1987), which won the Eileen Basker Memorial Prize of the Society for Medical Anthropology, she started to work on the anthropology of science and reproduction in the U.S., in particular on how gender stereotypes have shaped medical language. The next phase of her research focused on the interplay between scientific and popular conceptions of the immune system. In *Flexible Bodies: Tracking Immunity in America from the Days of Polio to the Age of AIDS* (Beacon Press, 1994), she analyses the manner in which the concept of 'flexibility' in immune discourse has been involved in a transformation of contemporary notions of health and business practices. Her present work is on theories of normalisation and the evolving constitution of selfhood in contemporary society.

Mary Morrison holds postgraduate qualifications in fine arts and critical care nursing. She has presented work on autopsy, pathology and critical illness utilising an interdisciplinary approach. Her current project has the working title of *Medicalised Portraits – Pictures of Individual Pathology*. She lives in Auckland (New Zealand).

Pamela Wood, a registered nurse, teaches nursing research at Otago Polytechnic, and recently has been seconded to work with staff across the Polytechnic in developing their teaching and research. She is currently a PhD candidate in the History Department, University of Otago (Dunedin, New Zealand), writing a cultural history of dirt, looking particularly at the relationship of ideas settlers held about dirt, settlement and public health.

Heather Worth is Senior Researcher at the New Zealand AIDS Foundation (Auckland, New Zealand), where she heads a number of social research projects: *Male Call*, a nation-wide study of men who have sex with men; *Couples*, a study of gay men in relationships; and a study of refugees in New Zealand and HIV. She is about to complete her doctoral thesis which is a critique of Michel Foucault's history of sexuality.

INDEX